How Big
and Still Beautiful?
Macro-Engineering
Revisited

AAAS Selected Symposia Series

 Published by Westview Press
5500 Central Avenue, Boulder, Colorado

for the

 American Association for the Advancement of Science
1776 Massachusetts Avenue, N.W., Washington, D.C.

How Big and Still Beautiful? Macro-Engineering Revisited

Edited by Frank P. Davidson,
C. Lawrence Meador and Robert Salkeld

Routledge
Taylor & Francis Group

LONDON AND NEW YORK

First published 1980 by Westview Press

Published 2018 by Routledge
52 Vanderbilt Avenue, New York, NY 10017
2 Park Square, Milton Park, Abingdon, Oxon OX14 4RN

Routledge is an imprint of the Taylor & Francis Group, an informa business

Library of Congress Cataloging in Publication Data
Main entry under title:
How big and still beautiful? macro-engineering revisited
    (AAAS selected symposium ; 40)
    Papers presented at a symposium held at the 1979
annual meeting of the American Association for the
Advancement of Science and sponsored by the Section on
Engineering and the Section on Industrial Science of
the AAAS, and the American Institute of Aeronautics
and Astronautics.
    Bibliography: p.
    Includes index.
    1. Engineering--Congresses.   I.   Davidson, Frank
Paul, 1918–   II.   Meador, C. Lawrence
III.   Salkeld, Robert, 1932–   IV.   American
Association for the Advancement of Science. Section on
Engineering.   V.   American Association for the Advance-
ment of Science.   Section on Industrial Science.
VI.   American Institute of Aeronautics and Astronautics.
VII.   Series: American Association for the Advancement
of Science.   AAAS selected symposium ; 40.
TA5.H67   620'.004'2   79-23498
ISBN 0-89158-792-6

ISBN 13: 978-0-367-02209-9 (hbk)
ISBN 13: 978-0-367-17196-4 (pbk)

About the Book

This volume examines fundamental issues of macro-engineering--now a planetary norm--from the viewpoints of psychiatry, social science, management, and law. The contributors suggest a general theory to guide future decisions on large-scale projects and programs and analyze specific cases in the context of a set of public-interest guidelines.

About the Series

The *AAAS Selected Symposia Series* was begun in 1977 to
provide a means for more permanently recording and more
widely disseminating some of the valuable material which is
discussed at the AAAS Annual National Meetings. The volumes
in this *Series* are based on symposia held at the Meetings
which address topics of current and continuing significance,
both within and among the sciences, and in the areas in which
science and technology impact on public policy. The *Series*
format is designed to provide for rapid dissemination of
information, so the papers are not typeset but are reproduced
directly from the camera-copy submitted by the authors, without copy editing. The papers are organized and edited by
the symposium arrangers who then become the editors of the
various volumes. Most papers published in this *Series* are
original contributions which have not been previously published, although in some cases additional papers from other
sources have been added by an editor to provide a more comprehensive view of a particular topic. Symposia may be reports of new research or reviews of established work, particularly work of an interdisciplinary nature, since the AAAS
Annual Meetings typically embrace the full range of the
sciences and their societal implications.

WILLIAM D. CAREY
Executive Officer
American Association for
the Advancement of Science

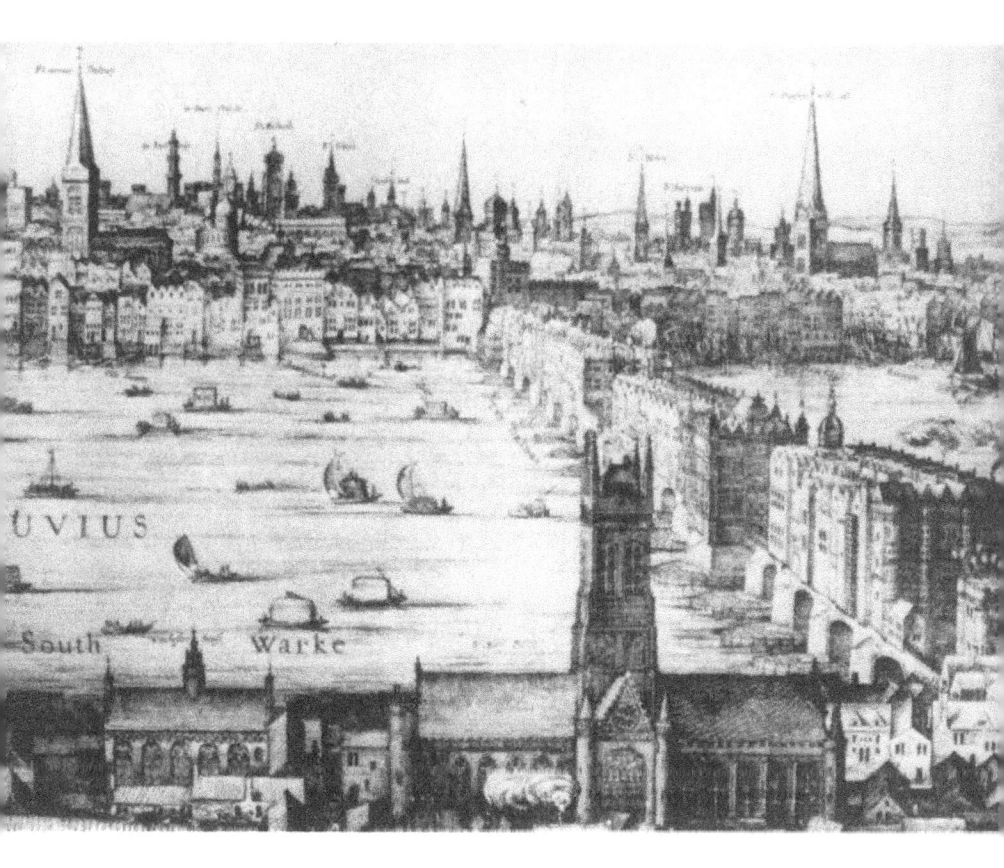

"London Bridge and the City Skyline Beyond." From a
lithograph by Wenceslaus Hollar, c. 1620. Courtesy
of the Trustees of the British Museum.

Contents

Figures, Tables and Illustrations

Chapter 15

About the Editors and Authors

Frank P. Davidson, *chairman of the System Dynamics Steering Committee, Alfred P. Sloan School of Management, Massachusetts Institute of Technology, holds a J.D. from Harvard Law School, but has been involved in macro-engineering for the past 20 years. He has been advisor on projects of unusual magnitude and complexity to the U.S., Canadian and British governments and to corporations and universities. He was founding president of the Institute for the Future in California and founder of the Channel Tunnel Study Group (planning group for the English Channel tunnel) and is a member of its governing board. He is also president of Technical Studies, Inc., the U.S. participant in the Channel Tunnel Study Group, and is vice-chairman of the Institute for Educational Services in Massachusetts. Among his many publications is* Macro-Engineering and the Infrastructure of Tomorrow *(AAAS Selected Symposium No. 23, Westview, 1978), of which he was senior editor.*

C. Lawrence Meador, *lecturer in the School of Engineering at Massachusetts Institute of Technology, is a specialist in computer-based information and decision support systems for engineering design, policy analysis, planning, and control. He is president of Decision Support Technology, Inc., and has served as consultant to the U.S. Departments of Energy and Defense and several major corporations. He has been assistant director, research associate and instructor for the Center for Information Systems Research, Alfred P. Sloan School of Management, and was a founding member of the Clinical Decision Making Group at MIT's Project Mac. He has published widely, especially on decision support systems, and is vice-chairman of the Eastern Hemisphere and Latin America Area Committee of the Institute for Electrical and Electronic Engineers' Computer Society. He also serves as an editor of* Computer Communications.

Robert Salkeld *is a consultant on aerospace and military planning to industrial and government organizations and chairman of the Technical Committee on Space Systems for the American Institute of Aeronautics and Astronautics. He participated in early studies of reusable launch vehicles and spacecraft and of space stations, and worked on Pioneer I, the first deep space probe. In addition to holding numerous patents on space systems devices, he is the author of more than 30 publications, including* War and Space *(Prentice-Hall, 1970). He is a fellow of several professional societies.*

Ramón Barquín, *manager for External Programs for IBM Americas/Far East Corporation and research fellow at MIT, is a specialist in technology transfer. He is the author of publications on transfer of computer technology, creation of technology in Latin America, and cultural differences and information systems utilization, and is chairman of the International Chapters Area, IEEE Computer Society.*

Kenneth W. Billman *is group leader of the Advanced Projects Group in the Materials and Physical Sciences Branch at NASA's Ames Research Center, specializing in space power concepts, advanced energy systems and laser energy conversion. He is the author of numerous publications on these topics, and is the editor of* Radiation Conversion in Space, *Volume 61 of the AIAA Progress in Astronautics and Aeronautics Series. He has been chairman of three national conferences on radiation energy conversion sponsored by NASA, and he holds three patents on his work in these fields.*

Stuart W. Bowen *is an aerospace engineer with Beam Engineering in California. His past areas of research have included design development and operation of DC arc plasma jets, quantitative plasma diagnostics, kinetics of gas and plasma reactions, and modeling and calculating recombination and excited state distributions in plasma flows and plasmadynamic lasers. His current research is concerned with calculation of characteristics of insolation on ground sites from a system of orbiting mirrors (SOLARES) and the design of the reflecting mirrors.*

John P. Craven *is currently Hawaii State Marine Affairs Coordinator, dean of Marine Programs and professor of Ocean Engineering at the University of Hawaii, and director of the Law of the Sea Institute. He has published widely in the fields of marine hydrodynamics and technology, as well as on arms control and the law of the sea. He is a member of the advisory committee to the U.S. delegation to the Third United Nations Law of the Sea Conference, and is currently a member*

*of the National Academy of Engineering, serving on its Marine
Board and Ocean Policy Committee. He has received numerous
awards and honors for his work, including Distinguished
Civilian Service Awards from the Department of Defense and
the Navy.*

William P. Gilbreath, *a research scientist at Ames
Research Center, is currently working on concept development
and feasibility of large projects such as the generation,
storage and transmission of large amounts of energy in space.
He has carried out studies of the interaction of aerospace
materials with various environments, including sea water,
rocket propellants, high vacuum and various gaseous species.
He holds memberships in various professional and honorary
societies and has published a number of papers in his fields
of interest.*

Cynthia Oudejans Harris *is a psychiatrist for the
Gestalt Institute of Cleveland; a faculty member at Case
Western Reserve University, Department of Community Health;
and a consultant psychiatrist to the Ohio state hospital
system. Among her many interests is a concern for the impact
of the modern architectural environment on the people who must
live in it.*

Ewald Heer, *director of Autonomous Systems Technology
at the Jet Propulsion Laboratory, Pasadena, is also an
adjunct professor of industrial and systems engineering
and director of the Institute for Technoeconomic Systems
at the University of Southern California. A specialist in
development of automated and space systems, he played a
leading role in the planning of the NASA robotics program.
He has organized and chaired two international conferences
on remotely controlled and automated systems; has written
many papers and has edited two books in these fields.*

Hans R. Huessy, *professor of psychiatry at the
University of Vermont College of Medicine, is a specialist
in child and community psychiatry and is concerned with the
social impact of macro-engineering projects. He has published articles on social organization and mental health
and is the author of* **Mental Health with Limited Resources**
(Grune and Stratton, 1966).

Cordell W. Hull, *vice-president and treasurer for the
Bechtel Corporation, has degrees in both engineering and
law, and has extensive experience in international project
management, project financing, banking, engineering and*

*construction, and contract negotiations. He has numerous
published articles in the areas of project financing and
administration.*

William J. Jones, *a senior research associate at the
Energy Laboratory, Massachusetts Institute of Technology,
specializes in alternative energy technologies. He has been
a lecturer in physics at Harvard University and a staff
member of the U.S. Army Corps of Engineers. Among his
several publications is* Ecology and Environment *(with
Richard Wilson; Academic Press, 1974).*

George Kozmetsky, *a specialist in systems analysis,
organization theory, information technology, and portfolio
management, is dean of the Graduate School of Business of
the University of Texas-Austin. He has served on the U.S.
Commission on the National Data Center, NASA's Management
Advisory Panel, and several other commissions and panels.
He is the author of numerous publications on information
technology, technology transfer, industrial innovation, and
management of large-scale systems.*

Jeanne Krause *is an historian and writer whose major
interests are economics and science and technology. She
was director of communications for the First Boston
Corporation and is a former associate editor of* Fortune
*magazine. One of her research interests has been land spec-
ulation along the Erie Canal in the early 19th century and
the local impact of this macro-engineering project.*

Andrew C. Lemer, *a civil engineer and senior associate
with Alan M. Voorhees & Associates (Planning Research
Corporation), has worked on the physical, policy, and
financial planning and evaluation of a variety of projects
and programs in the United States and abroad. In addition
to planning Nigeria's new capital, he has been involved in
studies of major airport needs (Australia), problems of
national goods movement (Iran), and financing of river
crossings (Tidewater Virginia). His particular interest is
in the appropriate role of large projects in development,
as mechanisms of technology transfer, and in effecting
institutional and social change.*

James A. McHale, *a consultant for agricultural products
companies, has been a farmer for more than 30 years. He was
Secretary of Agriculture in Pennsylvania from 1971 to 1976
and later became a member of the Governor's Office of Planning.
His particular concern is land rebuilding and the present and
future impact of macro-agricultural systems on soil quality.*

Francis Morse *is associate professor of aerospace and mechanical engineering at Boston University. His current interest is the future of the rigid airship, and he has published articles on airships and airship propulsion in journals around the world. He is a member of the American Institute of Aeronautics and Astronautics and the Lighter-than-Air Society.*

Arthur C. Parthé, Jr., *division leader for Energy Programs Technical Management Support, Charles Stark Draper Laboratory, holds advanced degrees in both mechanical engineering and business administration. He is chairman of the MIT Venture Forum, teaches courses in management information systems and management planning at Rivier College, and is co-developer of macro-engineering courses at MIT. He has authored several papers on project management and the application of modern control theory to macro-engineering projects and is a co-developer of MACRO-PLANNER-- A System for Planning and Control of Macro-Engineering Projects at MIT.*

John A. Seeger, *associate professor of business administration, Northeastern University, has also been a lecturer on system, world, and urban dynamics at Northeastern University's Graduate School of Engineering. He was formerly administrative officer, System Dynamics Group, and assistant director, Division of Sponsored Research, at MIT, and has been an officer in various commercial enterprises. His current interests include simulation in policy analysis, managerial models, and organization design and the implications of the limits-to-growth debate.*

J. Peter Vajk *is a senior scientist at Science Applications, Inc., specializing in space technology assessments. He is the author of numerous articles in both technical and popular journals on the long-term implications of space industrialization and colonization on global socioeconomic development. He was a major contributor to the NASA study, "Space Industrialization: 1980 to 2010" and was a participant in the NASA-Ames Summer Study on Space Manufacturing with Non-Terrestrial Materials.*

Acknowledgments

The Editors gratefully acknowledge the initiative and counsel of Sections M (Engineering) and P (Industrial Science) of the American Association for the Advancement of Science, and of the American Institute for Aeronautics and Astronautics. As joint sponsors of the Symposium which led to this volume, they deserve the appreciation of both scholars and practitioners.

Kathryn Wolff and the staff of the AAAS Publications Office provided original suggestions on a multitude of matters of taste and detail. They were unfailingly generous with time and insights even (or especially) when under severe pressure because of a congestion of deadlines.

Beverly Bugos and Edward Marvin of the macro-engineering group at M.I.T. furnished valuable technical assistance in the preparation of portions of the manuscript.

In the launching of this new and burgeoning field of inquiry, special mention must be made of the early support of Ernst Weber and A. George Schillinger, and of the generous participation of A. Ranger Curran, Paul H. Nitze, and Ellis Scott in the first (1978) AAAS Symposium. We are delighted that Arthur Herschman and his talented staff have made arrangements for a further Macro-Engineering Symposium at the 1980 AAAS Annual Meeting to be held in San Francisco.

Cherie Wallett typed the entire manuscript, unified presentation styles of key words and phrases, and collated texts and drawings from twenty contributors with indefatigible patience, tact and good judgment.

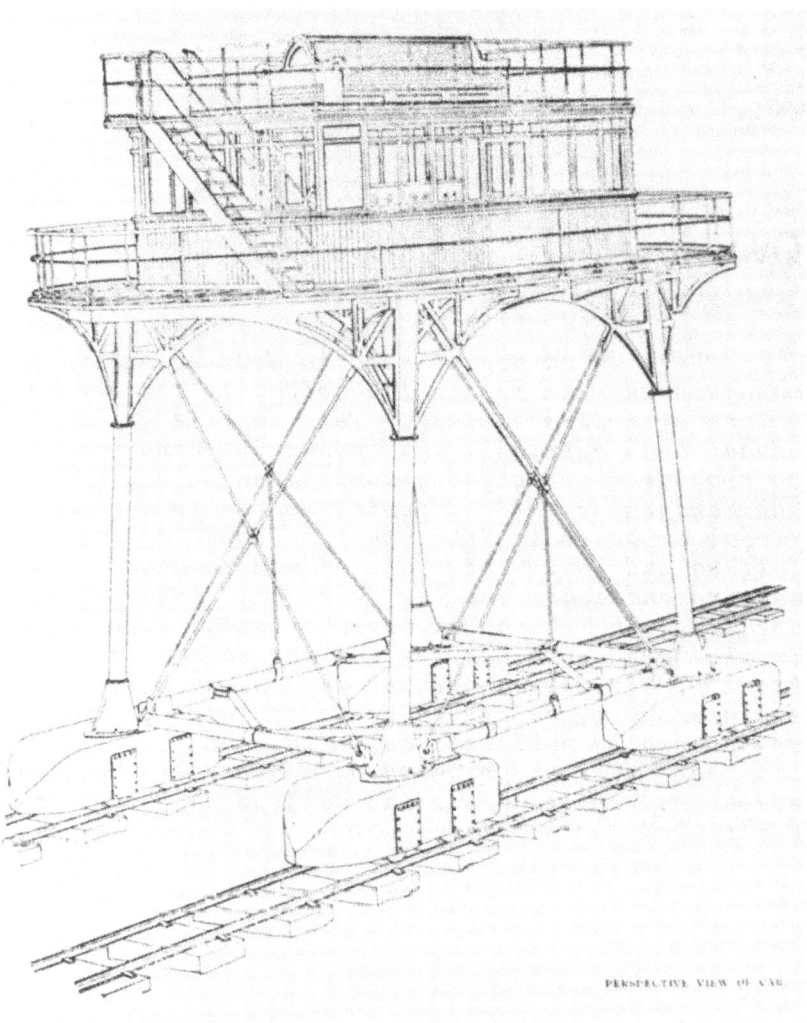

PERSPECTIVE VIEW OF CAR

The Brighton and Rottingdean Seashore Electric
Tramroad. Rails were laid directly on the sea-
bed and the passengers rode well above the waves
in a wagon raised on stilts above the axles.
The tramroad was described by R. St. George-Moore
in "A Through-Sea Railway," <u>Public Works</u>, Vol. 1,
July-October, 1903 (St. Brides Press, London).

Overview

There is an urgent need to develop engineer-managers who can relate technology to the deeper values of modern society. This implies broadened educational curricula and innovative career paths to provide thoroughgoing competence in engineering and management while offering more than a superficial exposure to "the other culture" of arts, letters and law. The scale of modern macro-engineering endeavors and the "external" costs associated with them have led to substantial government involvement, even in free enterprise societies. Frank P. Davidson concludes that because macro-engineering is so pervasive in its effects, it is an inherently political activity, and its practitioners must be prepared to deal sensitively with social and environmental as well as technical imperatives.

Frank P. Davidson

Introduction: Dilemmas of Scale and the Training of Future Engineer-Managers

How Big and Still Beautiful? Macro-Engineering Revisited has been contrived as a logical sequel to the first book in this field, Macro-Engineering and the Infrastructure of Tomorrow (1). Both volumes have been based on papers prepared for symposia at Annual Meetings of the American Association for the Advancement of Science. Last year's volume emphasized the ubiquity of large-scale infrastructure projects and the need for a better perspective on their impacts and management. The present work sets out, more aggressively, to shed multi-disciplinary light on the pre-suppositions, the semantics, the issues and the psychological, economic and political implications of humanity's awesome power to engineer the future of the planet.

The combination of engineering and organizational technology was a characteristic of early civilizations (2). What is novel in today's world is the enhanced precision and pervasiveness conferred on engineering by recent progress in the mathematical and physical sciences. Not surprisingly, the proliferation of macro-systems for power, transport and communications has been accompanied by a growing critique of the effects of industrialization on the social and physical environment. Harvey Brooks summarized this critique in a useful article, "Technology: Hope or Catastrophe?" (3), in which he reminds us that many writers "attribute our troubles not to science itself, but to its embodiment in a particular kind of technology, high technology, which is equated with large-scale, centralized systems, operated and controlled by large, impersonal bureaucracies, both corporate and governmental". He cites the paradox of

plentiful food and public hygiene - both products
of science and engineering - as root causes of the
population explosion and its consequences of under-
nourishment and disease! Thus regarded, science
and engineering, instead of offering solutions,
have become "part of the problem".

Writing a decade ago, E.V. Krick, in An
Introduction to Engineering and Engineering Design
(4), concluded that "analysis of the problem areas
in which engineering stands to make significant contributions reveals a special need for engineers who
bring to bear on problems a broader perspective,
more imagination, and a higher sense of purpose than
have prevailed up to now." Jay W. Forrester,
addressing the National Academy of Engineering, put
it more specifically: "...our educational system
is not designed to produce the special engineer-
manager with the high competence needed to integrate technology into our complex modern society."
(5)

In this respect, industry appears to have
jumped several steps ahead of academe: "engineer-
management" has already become big business and is
rapidly acquiring the status of a profession. Confronted with the task of planning and supervising
the construction of a new urban-industrial-port
complex with a price-tag that could exceed twenty
billion dollars, Bechtel Corporation was able to
draw upon a pool of experienced "engineer-
managers" developed within its own organization(6).
Meanwhile, engineering schools have, by and large,
deferred to schools of business and management to
provide a viable "career path" for engineering
graduates attracted by the challenge of macro-
system management. While business and management
students are routinely equipped with the rudiments
of computer technology, the graduates of engineering schools often have little exposure to management science and tend to join corporate and governmental bureaucracies at a menial level.

In fact, the preparation of cadres capable of
evaluating, planning, managing and communicating
about the macro-systems of the future involves
several distinct but inter-related issues. The
great sin attributed to "practical" applications of
science in the present century is their tendency to
sub-optimize, to over-achieve narrow and limited

goals at the expense of the larger systems on which
life ultimately depends. It is not enough, therefore, to bridge C.P. Snow's "two cultures" (the
sciences and the letters) unless both cultures
share an awareness of "the big picture" and its
underlying dynamics. This is a tall order; it
implies a more robust approach to General Education
than the token courses in physical science proposed by the literary deans of our great universities. We shall have to face an "agonizing
re-appraisal" of primary and secondary education
and ask ourselves, as Robert Wood has done, whether
it is not timely for private endowment, on a substantial scale, to rescue public school systems
from mediocrity wherever determination and leadership are present*.

Just as war has been judged "too important to
be left to the generals", so engineering has become
too pervasive to be left to the engineers: the
correct and confident use of technology is not only
in the public interest, it is essential to the
prosperity, the good spirit and even the future
safety of the republic.

The question of the training of engineers
would be less serious if it only involved the
career prospects of one species of graduate and the
economic health of the small group of private enterprises committed to the management of engineering.
The issue becomes one of urgent public concern
because large-scale technical programs absorb
increasing amounts of government investment. The
executive and legislative branches of federal,
state and local governments play critical roles in
the decision process, and of course the public
itself bears an inescapable responsibility for the
election - or defeat - of candidates for office.
It is essential, therefore, that a more coherent
and comprehensive education be provided for
"engineer-management", so that public agencies,
news media, research institutes, trade unions, and
business corporations may all have access to persons with the necessary "mix" of engineering,
biological, legal and management studies as well as
accompanying work experiences echelonned to

---

*Dr. Wood et al have organized a "Fund for the
Boston Public Schools" on the analogy of private
endowments for state-supported universities.

provide increasing scope and responsibility.

The public has a vital stake in the choice of the scale and the nature of future major technological developments. With government funding or government guarantees a frequent essential, the citizen's pocketbook is affected both by the impact on taxation and by the effect of physical facilities on the cost and quality of the delivery of various services. There is, moreover, the supervening issue of public health: in an era when mega-populations; widespread air, water and land pollution; and extensive <u>anomie</u> are fashionable topics, it becomes imperative to assess the indirect as well as the more easily foreseeable impacts of major investment decisions. Macro-engineering is rapidly becoming a planetary norm. What are the opportunities to employ our new, advanced technologies on behalf of individual, family and community health and well-being?

Historically, large-scale structures have had wide appeal because of their economy and the highly visible benefits resulting from such achievements as the Roman aqueducts, the Dutch dykes, and modern railway and communications systems. In certain societies there was a tradition and a <u>mystique</u> surrounding public works: in Rome, it was not unusual for wealthy citizens to build public baths or other facilities much as today's philanthropists endow research foundations and university chairs. In France there is still a residue of enthusiasm for "<u>les grands travaux</u>": Riquet de Bonrepos, the country gentleman who obtained a concession from Louis XIV to build a canal connecting the Atlantic Ocean and the Mediterranean Sea, remains something of a national hero. In Victorian England, civil engineering enjoyed a heyday, with Isambard Kingdom Brunel building the largest vessel ever launched to that time (1858) – the <u>Great</u> Eastern – and completing the first true underwater tunnel (under the Thames). It was, of course, a period of technological optimism. Today, it is seldom sufficient to establish technical and financial feasibility: the engineer must know in advance as much as he (or she) can about the impacts of a project on ecological factors, on people's perceptions and behavior, on energy supplies, and perhaps the international balance of payments.

In short, it is essential to understand the
system dynamics of a society where engineering and
politics are inextricably linked.

The risk of learning so much about so many
things is that we may soon become "a society of
evaluation" and lose the capacity for decision.
This need not be the outcome if we build a better
perspective on the central question of appropriate
scale: if all large enterprises are stymied or
torpedoed merely on the ground of size, we shall
clearly have arrived at a pathological point in our
history. Curiously, it is the late E.F. Schumacher
who best understood the dilemma. Had his well-
known work been provided with a different title, it
could have served as a basic text in macro-
engineering. In Small is Beautiful(7), he stated:

> "We always need both freedom and order.
> We need the freedom of lots and lots of
> small autonomous units and, at the same
> time, the orderliness of large-scale,
> possibly global, unity and coordination...
> What I wish to emphasize is the duality
> of the human requirement when it comes to
> the question of size: there is no single
> answer. For his different purposes, man
> needs many different structures, both
> small ones and large ones, some exclusive
> and some comprehensive.... What scale is
> appropriate? It depends on what we are
> trying to do."

For the engineering industry, the proliferation of gigantic projects has posed grave problems
of recruitment, training and management. The
Harvard Graduate School of Business Administration
has sponsored a series of thoughtful papers by
Professor Mel Horwitch and others on the dilemmas
of "Large-Scale, Public-Private, Technological
Enterprises".

> "The rise of such enterprises is a
> worldwide phenomenon... These endeavors
> possess both public and private compo-
> nents. They are also usually established
> to accomplish some mission, although such
> a mission can be as focused as landing on
> the moon or as ambiguous as implementing
> an overall energy program.... Society

confronts an increasing array of
large-scale, often technologically
innovative needs which are beyond
the technological, financial,
political, and managerial capabilities of any single organization."(8)

On the subject of the huge enterprises having
to do with the exploration and use of "outer space",
R.C. Seamans and F.I. Ordway published a landmark
report, "The Apollo Tradition: An Object Lesson
for the Management of Large-Scale Technological
Endeavors."(9).

The National Science Foundation, through its
Science and Technology Policy Office acting "in
support of the Director of NSF in his role as
Presidential Science Advisor" commissioned a major
study, "Constraints on Large-Scale Technological
Projects", whose Summary Problem Statement is worth
citing in some detail:

"Large-scale technological projects
(LSTPs), though never easy to achieve,
often have enjoyed overwhelming success
in the past. There is reason to believe,
however, that as the industrialized
societies mature it may become increasingly difficult to initiate LSTPs, or to
carry them through to successful completion. Ironically, this comes at a time
when the technical capacity to carry out
such projects has never been greater,
while the need to rely on them to deal
with rising social crises has never been
more urgent. Evidence for these difficulties in the form of technical failures,
long delays, abandoned projects, growing
regulation, and shifting public attitudes
continues to mount. The consequences
could be grave if major problems go
unsolved..."(10)

The United States, at this moment in its
history, faces a two-fold challenge. First, the
national slippage in productivity and competitiveness must be made good by a concerted effort of all
sectors of society. And second, we must demonstrate that our technological capacities can be
directed, wisely and effectively, on behalf of

human needs and aspirations. Both tasks require more than even the most rigorous analytical models, vital though these may be. We must look for such creative and adaptive energies as implied by Buckminster Fuller's familiar quest for "a comprehensive, anticipatory design science", and we must seek out the reinforcing medicines of integrity, persistence and a sense of public service. Macro-engineering is not a closet industry; there is an obvious and unavoidable political dimension.

For instance, the President and the Congress have an interesting opportunity to join with Canada and Mexico in cooperative planning for the development and conservation of the natural resources of the North American continent. Professor T.W. Kierans of Memorial University in Newfoundland has suggested one of many possible schemes for the mutually beneficial sharing of water resources: bearing in mind the dismal reception accorded the late Ralph M. Parsons when he proposed a North American Water and Power Alliance (NAWAPA) (11), it will be a test of diplomacy - and of our deeper understanding of the psychology and priorities of our neighbors - to explore such initiatives as Professor Kierans' with tact and thoroughness.

On another level, American government and industry would do well to ponder the report of the special Parliamentary Commission set up in London under Sir Montaque Finniston to explore all aspects of the British engineering industry: recruitment, training, investment, labor practices, marketing, etc. Our own country has not yet set in motion a comparable response to the erosion, frequently noted (12), of our trading position. To be sure, we have problems peculiar to our own history and traditions such as the confrontation legacy of the 19th Century anti-trust response (to practices which, at the time, merited stern resistance and regulation). Today it seems essential to design a new and more cooperative relationship among government, industry, labor and the universities. Necessarily, this broad re-grouping will call for initiatives that are inter-disciplinary, inter-professional and inter-sectoral. As the late Hubert H. Humphrey's foreign policy coordinator pointed out in the concluding chapter of Macro-Engineering and the Infrastructure of Tomorrow (13), macro-engineering can now make a substantial and positive contribution to both the domestic and the foreign policy of the United States.

I know of no better illustration of the symbiosis of large-scale concepts and organization with small-scale design and concern for the values of the individual and the family, than Peter Land's description, in Chapter Three of the earlier volume on <u>Macro-Engineering</u>, of "The United Nations Model Neighborhood in Peru". All over the world, the skyscraper-type of impersonal housing project has spawned disappointment and disruption. Peter Land has pointed the way, for developed and developing countries alike, to build low-cost individual clustered housing, attractively designed, as a viable and proven alternative to the monolithic "apartment house."

Such an approach, consonant with the latest findings of the behavioral sciences, may lead us to a better appreciation of the <u>psychological</u> importance of pleasant facilities – both urban and suburban – for active exercise and recreation. Percy Corbett's redesign of a Health Center proposed for New York City after World War II deserves careful study, as the engineering industry seeks ways to dramatize its potential for the service of a healthier and happier human future.

The 1979 Symposium in Houston, Texas, carrying the same title as the present volume, was fortunate to have had as its general chairman Dr. John E. Fobes, for many years Deputy Director-General of UNESCO. Dr. Fobes, perhaps more than any other individual, fostered international commitment to the re-forestation of denuded mountain regions. In countries such as Nepal, alternative fuel sources can surely be developed for mountain people, hitherto dependent on dwindling supplies of firewood, but whose patrimony includes several of the greatest untapped sources of hydro-electric power in the world. Here is an exemplary opportunity for creative statesmanship, for cooperative international efforts to reduce the devastating and recurrent floods which threaten the Gangetic plain. Macro-engineering, in this context, can make a deliberate and considered contribution to diplomacy and can assist in the improved "north-south" relationships and the more just international economic order which is so widely desired (14).

Responsible macro-engineering implies a comprehension of macro-economic factors, of the second-order effects of investments which are, both financially and politically, preclusive in character. The well-trained macro-engineer of the future will have a professional awareness of the leading methods of analysis, including system dynamics (15), decision support (16), and the new management techniques emerging from the broadening culture of schools of business administration (17).

Can there be, in the United States, a "career path" for managers and planners of macro-engineering ventures? Do we need a new combination of academic training and increasingly responsible practical experience along the lines of the centuries-old "ponts et chausées" system in France? Mere theoretical preparation, no matter how comprehensive, can only produce bright staff officers. Senior advisers and managers must be ripened on the tree of relevant experience. In a pluralistic society, this means that future leaders of engineering should have phased exposure to public sector, private sector and third sector responsibilities.

As for the academic arena, some useful combination of management, law, biology and engineering would seem to provide the best launching pad. A combined degree program covering undergraduate and graduate studies would be one approach worth looking at. Taking a leaf from the "cooperative agreement" long established as a mode of interagency cooperation in government, it should not be beyond the wit of democratic statesmanship to devise an agreed career path that will provide useful experiences with (as one possible grouping) the United States Army Corps of Engineers, a private engineer-management firm, a leading investment bank, a consortium of major building contractors and a powerful trade union. And as a counterweight to this purposive involvement with vast affairs, the candidate might well be required to experience the chores of a self-sufficient farm and of a small or intermediate industrial establishment!

Macro-engineering can best serve as a creative and adaptable instrument for achieving social and

commercial goals if its practitioners are individuals of high culture, deep learning and broad
experience. Sensitively tuned to personal and
community needs, with projects and programs selected by a sustainable consensus, macro-engineering
may help us re-discover the sense of a coherent
national and international mission.

The present volume is indicative of the wide-
ranging issues with which engineer-managers will
have to contend as an increasingly interconnected
world society approaches the year 2000. Topics and
viewpoints have been deliberately chosen to illustrate diversity: the "real world" is incurably
eclectic. However, in dividing the text into two
Parts ("Part I - Evaluating the Impacts" and
"Part II - Cases in Point"), the editors have tried
to suggest the need for both routes--the cognitive
and the experiential--in the exploration of socio-
technological complexity. On the one hand, attention must be given to models, conclusions, laws
and systems. In addition, there is the indispen-
sable corrective of specific cases, of life that
has been lived.

Chapter 1 discusses in a general way the
financial and contractual pre-requisites of the
gigantic mixed-economy enterprises so characteristic of our age. Cordell Hull, who spoke on this
subject to the participants in the Houston Symposium, brings to the table his practical experience
as a negotiator of macro-engineering ventures and
the special optique of the Treasurer of one of the
world's principal engineer-management firms. In
the following chapter, George Kozmetsky recommends
a more meticulous vocabulary in our discussions of
projects, programs and systems, and modulates the
lessons of his years of industrial experience with
the reflections resulting from sustained service as
Dean of a major school of business management.
William J. Jones approaches the task of evaluation
from a different perspective: that of a senior
research advisor concerned with the impact--or the
lack of impact--of grand designs on the persistent
problems of the deprived and the disadvantaged.
C. Lawrence Meador and Arthur C. Parthé, Jr.
conclude the opening section of Part I with an
exercise that employs the full panoply of systems
analysis and computer modelling: developing the
concept of a "macro-engineering state space", they

proceed to demonstrate a method of graphically
representing the multi-faceted sectors impacted by
large-scale engineering.

In a section entitled "Values, Psychology and
the Decision Process", Ramón Barquín leads off with
a series of questions about the linkages of large-
scale undertakings with human values. Cynthia
Oudejans Harris and Hans R. Huessy assess macro-
engineering from the special viewpoints of individual and social psychiatry. John A. Seeger, a young
academician with experience as an industrial consultant in the United Kingdom, focusses on the
decision process and, after summarizing the discussion to that point, urges the widest possible
participation on the part of the impacted public:
however financed or defined, major-engineering
ventures affect the public interest.

The second part of the volume begins with
Andrew C. Lemer's evocation of the proposal (with
whose configuration he has been directly involved)
for a new capital city in the interior of Nigeria.
John P. Craven, from his vantage-point in Hawaii,
tempts us with speculations on the use of maritime
potentials for the enhancement of urban life. And
James A. McHale, a practical farmer who became
Secretary of Agriculture of the state with the most
farms--and farmers--in the country, argues
passionately for a return to the family farm and
for the adoption of a national farm policy that
will put the health of the soil and of the rural
community ahead of commercial and industrial
values.

The editors have given transportation a
special section, not only because it accounts for
at least twenty percent of the gross national (or
world) product, but because throughout recorded
history all manner of transportation facilities
have provided a primary field for macro-engineering.
Jeanne Krause's retrospective commentary on the
Erie Canal reminds us of a time when, despite a
dearth of trained engineers, a single engineering
project helped catalyze the "take-off" of a young
and developing country. Francis Morse challenges
prevailing shibboleths and, despite the "curse of
the Hindenburg", sets forth a closely-reasoned case
for a whole new generation of super-dirigibles.
Finally, J. Peter Vajk treats us to a structured

look at alternative global transport systems which,
at the turn of the century, might make two-hour
trips between any two points on earth a commonplace event. If this is so, or even approximately
so, the word "togetherness" will have more than
psychological connotations.

The concluding chapters address issues of
space utilization and development. Kenneth W.
Billman and William P. Gilbreath with Stuart W.
Bowen provide a detailed assessment of orbiting
solar reflectors as a major world power source.
Not blinking at total system costs that could
eventually pass the trillion-dollar mark, I think
this proposal makes crystal-clear the underlying
links between macro-engineering and politics.
Ewald Heer's sober report on the economies and
advantages of robotization in outer space provides
a final impetus to reflection on ends and means,
on the allocation of necessarily limited resources
among competing visions of the future, on the
purposes and values which should predominate in a
human society no longer sure that it is confined
to its own biosphere.

In the midst of such quandaries, how is
humanity to make sensible investment decisions? On
a national basis, there are doubts as to how much
technology is appropriate for defense, how to
assure energy supplies without unacceptable hazards
to public health, and how to bring about effective
cooperation between government and industry to
maintain adequate productivity and competitiveness.
To resolve such issues, it would be helpful if
there were ready-made, so to speak, a General
Theory of Macro-Engineering. This volume may be
regarded, for all its diffuseness, as a modest
step in the direction of such a theory. If
engineering has truly become so sensitive and
central a topic that its parameters must be familiar to business managers, civil servants, politi-
cians and journalists, then the full inter-
disciplinary resources of our universities and
research institutes will have to be concentrated
on an effort to educate the public in the underlying dynamics. We shall have to put behind us
the naive assumption that there is an inescapable
conflict between environmental protection and
technological progress: the problem is one of
defining more precisely the environmental goals we

wish to attain and then designing the engineering
and institutional systems that can appropriately
support our environmental and social concepts.
That there must be "trade-offs" along the way, few
can doubt. But the calculation of cost-benefits
which include "external" impacts is no longer a
novelty. If some new institutions are needed, such
as a re-conceptualized Civilian Conservation Corps
to reinforce public efforts at soil conservation
and forest development, the public will surely be
ready to consider proposals once a logical and
prudent framework has been devised and presented.

Congress called attention to this general
problem area when it set up the Office of Technology Assessment. More recently, the President of
France established, by a special decree, the
<u>Institut Auguste Comte</u> with the principal mission
of training managers to assess and guide "major
programs of equipment". In the United States, it
may soon be time for a further step, beyond the
tentative studies of the Office of Technology
Assessment and well beyond the crisis-oriented
response to unanticipated dilemmas. Just as
macro-economics provided a more comprehensive view
of economic phenomena, so we may expect that a
long-range, comparative, inter-disciplinary
analysis of macro-engineering programs and systems
will make an important and perhaps at times a
decisive contribution to public policy.

References

(1)   Westview Press, Boulder, Colorado, 1978,
(AAAS Selected Symposium 23, edited by Frank P.
Davidson, L.J. Giacoletto and Robert Salkeld).

(2)   Eugene S. Ferguson in ibid., Chapter 1,
"Historical Perspectives on Macro-Engineering
Projects". Yigael Yadin, in The Art of Warfare in
Biblical Lands - in the Light of Archaeological
Study, McGraw-Hill, New York, 1963, pp. 32-35,
describes the discovery, in 1954, of extensive
Neolithic fortifications at Jericho "built some
6000 years before the time of Joshua." Yadin
makes it clear that macro-engineering has been with
us for at least nine thousand years.

(3)   Technology in Society, Volume 1, Number 1,
Spring 1979, pp. 3-17. Pergamon Press, New York.

(4)   John Wiley & Sons, Inc., New York, 1969,
p. 200.

(5)   "Engineering Education and Engineering
Practice", in Engineering for the Benefit of Man-
kind, pp. 129-144, National Academy of Engineering,
Washington, D.C., 1970.

(6)   Richard P. Godwin, "Planning and Conduct of
Macro-Engineering Projects", Chapter 4 in
Macro-Engineering and the Infrastructure of
Tomorrow, supra.

(7)   Harper Torchbook, New York, 1973, pp. 61-62.

(8)   "Uncontrolled Growth and Unfocused Growth:
Unsuccessful Life Cycles of Large-Scale, Public-
Private, Technological Enterprises" with special
reference to the United States SST Program and the
United States Attempt to Develop Synthetic Fuels
from Coal, presented to Symposium on the Management of Science and Technology, held in
Rio de Janeiro, June, 1978, sponsored by COPPEAD of
the Federal University of Rio de Janeiro and by
IA/FEA of the University of Sao Paulo. The quota-
tions in the text are from pages 2 and 3 of this
paper.

(9) Interdisciplinary Science Reviews, Volume 2, Number 4, Heyden & Son Ltd., London, 1977, pp. 270-305.

(10) From III-1, Assessment of Future National and International Problem Areas, Volume II, February 1977, (Reprinted September 1978), by Norman B. McEachron and Peter J. Teige with contributions by Harold A. Linstone, prepared for National Science Foundation, Contract NSF/STP76-02573, by SRI International, 333 Ravenswood Avenue, Menlo Park, California 94025.

(11) The Ralph M. Parsons Company, Los Angeles, California, 1966.

(12) See for instance The Future of Science and Technology, Victor J. Danilov, editor, Museum of Science and Industry, Chicago, Illinois, 1975.

(13) David Fromkin, Chapter 14, "Macro-Engineering as an Element of U.S. Domestic and Foreign Policy."

(14) Laszlo, Baker, Eisenberg and Raman, The Objectives of the New International Economic Order, published for UNITAR by the Pergamon Press, New York, 1978.

(15) See Jay W. Forrester, "Changing Economic Patterns", Technology Review, Cambridge, Mass., August/September 1978.

(16) Peter W. Keen and Michael S. Scott Morton, Decision Support Systems, Addison-Wesley, Reading, Mass., 1978.

(17) See for instance William H. Gruber and John S. Niles, The New Management, McGraw-Hill, New York, 1976.

Part 1

Evaluating the Impacts

Finance, Management and Control

Financing the Project

Section 3

Finance, Management and Control

Ferdinand de Lesseps, who in 1858 founded the
company responsible for building the Suez Canal.
The Canal reduces the sea distance between Western
Europe and India by nearly 5,000 miles. (Reprinted
from Vanity Fair, November 27, 1869.)

Overview

Macro-engineering projects are becoming larger, costlier and more complex; are subject to increasing risks and uncertainties because of worldwide economic and social dynamics and frequently require the capital support of national governments and/or international organizations. Although many macro-engineering programs are being cancelled, postponed or refused, Cordell W. Hull points out that the demand has never been so evident for projects which can supply the resources and infra-structures of the future. New techniques are emerging to design the financial sponsorship of large-scale technological ventures so that risks are widely and equitably shared. Much effort must be devoted to the analysis of contingencies and impacts in order to satisfy the sophisticated exigencies of the providers of "macro-finance."

1. Macro-Projects and Macro-Finance

Introduction

In the past few years, there has been an immense growth in the size and complexity of major industrial, energy, infrastructure and natural resource projects. Increasing costs, growing economies of scale, declining capital productivity, worldwide inflation, less economically attainable resources, higher developing-country indebtedness, remote locations, new technologies, supporting infrastructure requirements, expanding government regulations and environmental concerns are serving to make the large-size project increasingly difficult to develop and finance.

The undertaking of such macro-projects requires massive resources and involves numerous parties, all of which must be closely coordinated in the development, execution and operational phases of each such project. This includes the marshalling of many resources and integrating into the project, on appropriate terms, equipment and raw material suppliers; investor-owners; government export credit agencies; international, commercial and development banking institutions; shippers of equipment, raw material and final products; engineering/constructors; and purchasers of the project's end products. Increasingly, the host-government for the project is involved as an owner, or a guarantor of project debt or, at the very least, in a multitude of regulatory aspects; often it is the owner of the natural resources connected therewith. Frequently these macro-projects cut across national boundaries, requiring international assistance and cooperation from disparate sources.

The period from gestation of the project concept through planning, engineering and construction to operational start-up generally extends many years and often is a decade or more in today's world. Consequently, development and funding of these projects require consideration of the interests and concerns of the diverse participants, forecasting of changing economic, technological, market and resource implications over a prolonged time frame, and an appreciation of the impact of powerful geopolitical, social and environmental forces.

In order to undertake such a macro-project, all aspects must be carefully assessed and planned in advance. This involves extensive feasibility evaluation, obtaining commitments of natural resources and marketing outlets, engineering studies, and regulatory compliances. Project partners must be located, funding commitments sought, and negotiations undertaken with the concerned government agencies. Development and carrying costs to time of start-up will be very high and the various costs of delay can quickly impinge upon the project's economic viability.

This paper examines some of the major facets of funding and development that impact upon macro-projects in the context of today's international environment. These include economic, technical, financial, governmental, and management concerns. The planning and provision for these concerns determine the ultimate feasibility and success of the macro-project.

Current Economic Issues

Increasing costs, worldwide inflation, the growing scarcity of energy and natural resources, government regulations, the pace of technological innovation, currency instability, capital funding and allocation problems, and many other factors have combined to create a complex economic environment in which to undertake a project.

During the ten-year or so long period of development required, about which I spoke previously, what will be the inflation rate impact on costs of labor and materials for developing the project--five to ten percent or more per year--for materials and services purchased in the industrialized

countries? Make an estimate--there can be an
error of 100 percent or more in project cost if
the estimate is wrong. Further, price inflation in
the developing host country where local material
and labor may have to be obtained can range from
25% to 75% and up per year. These rates result in
a dramatic compounding impact on project costs over
the period of years we are considering. It is
probably reasonable to assume that inflation will
continue to increase capital project costs mater-
ially over the next ten years. Furthermore, the
impact of interest charges on loan funds used for
development and construction rapidly mounts over
the extended period. For the macro-project, the
interest during construction and the escalation
costs combined will often equal or exceed the
initial estimated basic project cost. In addition
to the normal prolonged period of development
which may be expected with the macro-project,
further delays can sometimes result from management inefficiencies, financing difficulties,
engineering, procurement and construction problems,
changes in marketing conditions, governmental intervention or regulations, and <u>force majeure</u> occur-
ences.

     The nature of costs, engendered by delays,
depends upon the stage of development (planning,
engineering, construction or start-up) affected
by the delay. Delays occurring at the outset of a
typical project impact costs primarily because of
the escalation in the price of materials, labor
and other requirements--such as insurance and taxes.
As project delays occur during the latter stages of
development, the associated costs are more a
result of accruing direct capital costs such as
interest and financing fees.

     A major indirect cost is the loss of return on
invested equity resulting from the delay in operations and the foregone generation of revenues and
income. Where a finite resource is being developed
over, say, a 10-30 year period, this return may not
be lost but rather delayed. However, after giving
effect to the time value of this return, its
discounted or present value may be greatly diminished.

     There is another type of indirect cost that is
important but often overlooked. Certain project

payments such as property taxes, income taxes or other taxes, various types of royalties and license fees, and other cash flows to the host government will be delayed or lost if the project is slowed. This is particularly important where a major resource development is involved. The total amount of all such added delay costs can exceed $200 million for each year of delay for a $1 billion macro-project. So, lost time is devastating to project economics!

Examples of rapidly escalating costs which serve to decrease real production capacity per construction dollar unit cost in macro-projects abound.

The Asahan hydro-power and aluminum project in Sumatra, Indonesia, being sponsored by Japanese-Indonesian interests, has increased in cost almost five-fold since 1972 ($420 million to over $2 billion).

A coal gasification project of 250,000 c.f. per day capacity costs $1 billion today, over double from 1973 and the gas production cost connected therewith is now estimated at $3.00 per 1000 c.f. compared with $1.32 when it was in the planning stages.

It costs a billion dollars to open a 150,000 ton per year copper mine, in a developing country, including infrastructure--triple its former cost in 1973.

So, in the face of the various uncertainties, capital hesitates. It turns to less indeterminate outlets and seeks quicker returns.

Yet supply demands, which the macro-project will need to service, continue to mount. Worldwide, in the mining area, much new capacity will be needed between now and 1990 for the development of iron ore, bauxite, alumina, nickel and other metals.

Growth in the demand for many such natural resources could increase four to five times through the end of the century. Worldwide, the energy industry will require many billions of dollars for oil and gas developments, nuclear plants, coal projects and the new emerging technologies in the solar, geothermal and synthetic fuel areas.

Will these projects be developed when needed?
If not, the impact upon the world economy will be
substantial. Higher inflation and lower GNP
growth and higher unemployment will be the inevitable result.

Turning to support for these projects from
various areas of the world, we discern that
European countries are reducing direct investment
in macro-projects in both Europe and the developing
world due to a chronic slowdown in economic growth.
However, this adverse trend is perhaps more than
offset by increasing amounts of debt financing made
available to such projects on preferential terms
through the medium of export credits supporting the
sale of material and services for the development
of such projects.

In the United States, macro-projects in almost
every sector are continuously being delayed because
of:

- Economic uncertainty
- Capital unavailability
- Lack of an energy program
- Uncertainty on future energy and natural
  resource demand and prices
- Increased government regulations and
  environmental controls
- Questions on nuclear energy
- Unresponsive Public Utility Commission
  attitudes concerning appropriate rate of
  returns, allowances for work in progress
  in rate bases, etc. for new major power
  projects
- Inadequate programs for export support

Further, because of various uncertainties
including the changing nature and philosophies of
developing countries toward ownership and development of their natural resources, the flow of
investment capital from the United States for the
development of macro-projects in the Third World
has faltered greatly in recent years. However, it
does appear that debt funding for such projects
insofar as they use U.S. goods and services is now
becoming more available from the U.S. Export-
Import Bank.

In the last decade, Japan's rapid economic growth supported the development of many major energy and natural resource projects in Third World countries. While direct investments in such projects have dropped drastically, Japan's aggressive export program, supported by advantageous export credits, is still providing very considerable funding impetus to the development of international macro-projects. Also, Japan is currently considering increasing its foreign investment program.

All of these types of factors affect the project's economics and funding availability and emphasize the need for detailed master and feasibility planning, and in-depth economic studies of the project, the markets and the forward longer term economic environment into which the project must fit. Increasingly, efforts must be made to reduce project delays and, as never before, the most professional talent available for macro-project development and implementation must be brought to bear throughout the feasibility, planning, financing, engineering, construction and operational phases.

Capital and Funding

Where can the enormous sums required for funding the macro-projects be located?

In the recent past, projects could be funded, to a large degree, directly by the sponsors from a mix of capital reserves and current cash flow. These funds were supplemented by debt through banks or institutional lenders, usually guaranteed by the sponsoring corporation. However, due to the increase in scope, cost, risk, and number of projects, owners are unable to follow such a traditional financial approach, certainly with respect to the macro-project. Instead, much of the required funding support for such projects is in the form of debt from traditional capital markets, eurodollar sources, or governments. Further, governments aid project development by guaranteeing loans and by allocating funds in the form of grants, concessional lending, export credits and investment guarantees. A macro-project is generally so capital intensive that it must be financed from a variety of such sources, including inter-

national commercial banks, eurodollar lenders,
local currency sources, institutional lenders in
the United States or Europe, export credit agencies,
the World Bank, development banks and the like.

Putting together an appropriate funding package from such diverse institutions requires an
intimate knowledge of the world's financial markets, an appreciation of lenders' concerns,
requirements and preferences, as well as creative
and responsive techniques in structuring the elements of a project to qualify for the type of loan
which meets with the project's amortization ability.

As such projects increase in size, both in
absolute terms and in relation to the capital of
potential corporate participants, the trend toward
consortia and joint ventures of multi-sponsors in
developing and investing in such projects may be
expected to continue. For example, this may take
the form of a $500 million copper project such as
Bouganville where a multinational group of companies and banks from the U.S., Canada, the U.K.,
Japan and Australia joined together to develop and
finance the production of 160,000 tons per year of
copper contained in concentrate; and it may also
take the form of a $3 billion integrated steel
mill, such as Aconimas in Brazil where a multinational group of companies from the U.S., Europe
and Brazil joined together to develop an installation to produce two million tons of ingots per
year, plus 1.5 million tons of products, and it may
also include direct ownership participation by the
host government, as also is the case with Acominas.

This approach has many advantages. First, it
serves to broaden the sources of equity and
increase the amounts available for this purpose.
Second, it spreads the risks of project completion
and operation, and provides more sources of additional funds to cover any project cost overruns.
Third, it would seem to lessen the risk of arbitrary actions, such as expropriation by host
governments, since such actions would impact
nationals from various countries, all of which
ostensibly may be affronted by such actions.
Finally, it provides a broader base of development
and operational expertise, diversifies the marketing base and outlets for the project's product and
increases the sources of export credit assistance

and bank funding for development of the project.

As to the sources of loans, they continue to proliferate. Liquidity in the eurodollar market is being exponentially expanded by such factors as the United States' balance of payments deficits, recycling of Arab petrodollars, the multiplier effect of fewer reserve or capital requirements of banks operating in this market, and continuous repayments of outstanding eurodollar loans. Many of these eurodollars are available for loans or investments in macro-projects.

The balance of payment problems of the United States and stagnated local economic growth in Europe and Japan have increased amounts of export credits available for macro-projects to take advantage of the continued absorptive power of international projects for exported goods and services. New sources of funds are opening up such as from the rapidly evolving financial markets of the Arab world as well as in the form of government credits supporting the export of services and materials from developing countries, such as South Korea, Brazil and India. Also, since the debt of developing countries is high (over $200 billion), commercial banks are increasingly directing their attention to feasible foreign exchange generating projects whose integral commercial viability and internal cash flow can support additional loans, permitting direct or indirect funding for activities within developing countries already heavily indebted to continue selectively. Furthermore, through participating in such projects and becoming involved in financial planning, letters of credit, fund transfers, performance bonds, interim contractor financing and the like, banks can greatly increase their own returns over instances where they simply participate as lenders in thinly margined balance of payment loans.

Major equipment, raw material or prospective market product offtakers from the project are increasingly willing to make loans or advance payments on projects to secure markets or sell their products. The World Bank is increasing its lending in the energy sector. So far the institution has made over 300 loans totalling over $9 billion for power projects and by 1990 such loans are expected to reach upwards to $1 billion

a year. Since the Bank only lends a portion of the
project's costs, this could indirectly generate
additional cofinancing from other sources exceeding $3 billion a year. Loan packages which involve
parallel or joint arrangements with the World Bank
also provide the advantageous involvement of the
World Bank's highly qualified staff in assessing
the project and helping monitor loan drawdown and
project progress.

Funding Source Concerns

Meeting the requirements of the financial
institutions requires a financially sound credit
and a well conceived project, with provisions for
timely cash flow generation, adequate returns for
investors, appropriate debt coverage ratios and
adequate security for the debt. Each funding
institution connected with the project will demand
a financially sound credit.

Some of these sources only provide medium-term
loans while others, such as bond markets and
export credit agencies, provide longer term funds.
The various sources of funds must be appropriately
blended so that the project cash flow, currency of
repayment, loan repayment tenures, form of security
and interest rate structure satisfy the requirements of the project. This presents an extremely
difficult task in connection with the macro-projects
for several reasons.

First, there is a question as to loan security.
Rarely are the sponsors of the macro-project of
sufficient size that they could absorb the full
impact of the debt on their balance sheets. So,
some form of off-balance-sheet project financing is
usually necessary and the lenders' prime security
increasingly depends on the cash flow and commercial viability of the project instead of the more
diversified and broader resources of an already
established business. This immediately raises the
question as to how risks and benefits should be
allocated among the owners of the project, the host
government, the market offtakers, the supporting
engineer/contractors, and the other involved
parties as well as how their contractual and economic relationships should interrelate. In
assessing the credit and the probability of loan
repayment, the lenders must be acquainted with the

raw material arrangements, economics, capital costs, marketing arrangements, and cash flow of the project and be satisfied that they are adequately and appropriately covered for cost overruns which may be incurred in completing the project, late completion, failure to complete, post completion shut-down, governmental interventions, expropriations or complete project failure. The macroproject sponsors must present a complete and detailed evaluation of each of these risks, and outline precisely how they are to be covered. Generally speaking, this will often result in the project sponsors' unconditionally guaranteeing completion of the project and committing for any additional funds which may be required for that purpose. In addition, they may have to agree to fund working capital or funding deficiencies for stated purposes up to agreed upon amounts during the actual operation of the project itself. Guarantees from the host government are frequently sought or obtained for a portion of or all of these types of risks, either because it has a vested direct interest in the project or because economic benefits in the form of taxes, duties, production of foreign exchange, development of resources, and training and employment of nationals will result from the project.

Regardless of such interim guarantees and assurances, it is fair to anticipate that once the project has commenced production operations, has proven its viability and sound marketing commitments have been arranged, that certain of the initial security undertakings may be allowed to fall away and lenders may be more willing to look to the security of the project itself including the cash flow being generated. This is particularly so if feedstock is assured at a given pricing formula, operating costs have been established and long term contractual arrangements are in place insuring marketing of the product at profitable margins during the period of debt repayment.

Lenders are rapidly increasing their capability to understand and perceive project risks. Accordingly, they are rightly insisting on more detailed feasibility, economic, and engineering studies of all the risk variables. The more thoroughly these elements are presented and sensitivity factors developed, showing the effects of changes in costs of construction, revenues, interest rates and operating costs on returns on equity and ability to repay debt, and appropriate arrangements demonstrated for the involvement of qualified, experienced firms in studying, developing and operating the project, the more readily an appropriate funding package will fall into place. Increasingly, I predict that lenders will insist on higher compensation for their money and services as they inevitably assume some of the risks themselves which are a part and parcel of major projects.

A most important factor in project evaluation is the market projection for the project product when it starts up, ten years or so out. What will the world economy look like then? Will there be a prolonged period of slow economic growth? Will market conditions have changed? Will there be product substitutions or technological break-throughs rendering the product obsolete or changing the project's economics? Will there still be a hospitable host government?

For the macro-project these are exceedingly difficult but imperative questions. Because of the long period for developing these projects and the uncertainties in the economic environment, lenders have become increasingly wary about the possibility of cost increases, delays in completion and the ability of the project to market its products far in the future to produce adequate revenues, leaving appropriate margins of safety and cash flow available for debt service. Accordingly, increasing attention must be given to working with lenders so that each of these risk areas is thoroughly studied and understood by all the parties concerned with the project, that all reasonable risks have been considered and allocated among the parties and elements of the project, and adequate assurances have been provided that either the risk is remote or that it is adequately secured against.

In today's macro-projects, an adequate rate of return must be allowed in the initial projections to provide a reasonable residual rate of return to the project's sponsors, provide a satisfactory margin of coverage for the debt and allow for various contingencies which the macro-

project will inevitably encounter, considering the
indefinite economic situation in the future when
the macro-project will come on stream, the immense
size of the project and the very considerable time
including both anticipated and unanticipated delays,
which will occur prior to completion. Also, the
unknown applicable inflation rates and other cost
factors of the project, the currency exchange risks,
and potential local host country government risk
because of indefinite and unknown prospective
policies within the host country bear directly on
the required returns. Whether macro-projects are
really being structured to provide appropriate
returns, taking into account all of the various
risk elements is indeed questionable and is
another area of increasing concern to the lenders
and to the investors seeking positions in such
projects.

Turning to the currency exchange risk question, because of the necessity to blend together so
many sources of funds for the project, an international macro-project often ends up with debt
commitments in at least several and often many
different currencies. While it is impossible to
project with any degree of certainty over such a
long time, the cumulative impact of currency
movements on the project, it can be taken for
granted that currencies will move with respect to
one another over the time frame in question and
that such movements will have a most substantial
impact on the economics and debt servicing capabilities of the project. Without attempting to
forecast precisely what this impact might be, it
is good practice to analyze various possible
ranges of currency movements with respect to the
currencies in question and determine how, at least
over the period the debt is outstanding, they will
impact on the project's economics. Often adverse
movements will be shown to erode substantially into
viability and may militate against committing for
debt in certain currencies, or result in a preference of one source of loan over another. Ideally,
the currencies of debt should be matched as nearly
as possible to the currencies of revenues which
will inure to the project during operation. Since
this is only infrequently possible, at least
attempts should be made to maximize this feature
while committing the remainder of debt in those
currencies whose movement might be expected to move

or remain more closely in line with the movement
of the currency of revenues of the project--a
complicated, frustrating and sometimes non-produc-
tive analytical task, but one that really should
be done.

Other Issues

There are many other issues which I can only
briefly touch upon here. With respect to macro-
projects in developing countries, most governments
are becoming increasingly involved either as a
major co-sponsor or as sole owners, or at least
guarantors of the debt, particularly in the cases
of natural resource development projects.

Here, questions arise such as where will the
expertise be obtained for evaluating, planning,
managing, constructing and operating the project.
Often personnel experienced in these types of
endeavors are in short supply in developing
countries. Where will the product be marketed
since the country may not be effectively integrated into an effective distribution system?

The country may have broader considerations
transcending project economics, i.e., national
security of supply, employment, prestige, social
goals, import substitutions, a wasting natural
resource (flaring of gas), etc., which require
the implementation of the project in spite of what
may be marginal or questionable traditional project economics. How should such a project be
structured and presented in competing for international capital funds?

As the periods of time for project implementation are prolonged, how is equity capital to be
directed to such projects in view of quicker payout
project opportunities available? The returns must
be sufficient to attract such capital, and tax and
other investment incentives should be considered
by all concerned governments to enhance the
project's viability.

What are the future implications to the
country and to the world economies arising from the
failure of governments to provide a hospitable
investment environment, of governments failing to
appreciate the vast risks inherent in such projects,

failing to allow the returns necessary to attract
capital to the macro-project, or failing to provide long-term regulatory certainty? These
questions are directed as much to the industrialized world, in my case to the U.S., as to
developing countries.

Conclusions

So, drawing several observations out of this
discussion in relationship to development and
funding of macro projects, I would submit for
consideration the following:

1. Macro-projects are getting bigger, more complex
and costly, and take increasingly longer development time.

2. The number of parties and diversity of nation-
alities connected with a macro-project is growing.

3. Governments are playing an increasing role in
these projects.

4. They cannot be viewed in isolation and they are
increasingly tied to world economic and geopolitical forces.

5. The economic uncertainty of the world affects
the timing of and type of which macro-project
proceeds.

6. Because of economic uncertainties, these types
of projects are being increasingly deferred.

7. The cost of delays, which are most substantial,
can amount to as much as 20% or more per year of
the initial estimated capital cost.

8. Delays in the construction/development of
macro-projects in the energy, natural resource and
other essential sectors so necessary to future
world economy will inevitably mean raw material and
energy shortages sometime in the near future.

9. The result of this impact will be inflationary
and depressing to economies and employment.

10. When developing a macro-project, increasing
attention should be given to the feasibility

analysis, including sensitivity analysis of
variations in the essential parameters of capital
cost, revenues, cost of money, operating cost and
the like, and professional studies must be made of
the impact upon the project's economics and the
ability to repay debt of:

- fluctuations in currency exchange rates of the
  various currencies in which the debt is
  committed and the currencies and revenue of the
  project;
- cost of delays;
- prospective markets;
- the international economic environment;
- the program for project implementation.

11. Increasingly, funds are sought for project
development from other than internally generated or
guaranteed sources. Therefore, projects must be
structured so as to qualify for loans from numerous
sources on a project financing basis.

12. While the sources of capital for macro-projects are increasing and becoming more diverse,
capital shuns uncertainty and will be allocated
where the risk-return aspects are most favorable.

13. Lenders increasingly are concerned with the
risk factors such as cost of completion, delays,
cost overruns, viability of the marketing arrangements, technical appropriateness of the project,
and the manner in which the project will be managed
during the construction and operational periods.
Funds will flow only to those macro-projects which
are conceived to take into account the issues which
have been discussed in this paper and which have
made a sound and responsive evaluation of the
various risks connected with the project, have
provided appropriate answers and security mechanisms with respect to those risks and have arranged
for qualified and experienced personnel to study
and manage the development, engineering, construction and operation of the project.

14. New techniques of structuring sponsorship of
projects must be sought including novel arrangements amongst the equipment suppliers, engineer/
contractors, owners of the natural resources,
traditional sponsors, host governments, and market
offtakers to expand the equity funding base and

provide a complete integral package of all skills
and services needed to make the project a success
and spread the risks while assuring each such
participant an adequate return.

Note

This paper was also delivered at the Asian International Chemical and Process Engineering and
Contracting Show in Singapore, January 16, 1979.

Overview

It is appropriate to develop new tools and
techniques for the evaluation and management of
macro-engineering efforts because the macro-systems
are no longer self-contained but are highly inter-
related. George A. Kozmetsky argues that traditional conceptual models for management are limited
insofar as they are restricted to single objectives
of an economic nature, to self-contained operational units, and to static environments. Impetus
for the development of macro-systems management can
logically come from the senior executives and administrators of our major institutions: these are
the individuals who bear the operational responsibility for assessing policies, programs and plans.
These executives have an opportunity to develop the
new instrumentalities for inter-sectoral policy
analysis and for the management strategies needed
to maintain a viable society.

2. Evaluation of Macro-Systems: Models and Case Analysis

The occasion of the January 1979 American Association for the Advancement of Science meeting is an appropriate setting to revisit macro-engineering and to evaluate its future as to "How Big and Still Beautiful." One of my early findings in revisiting the literature on macro-engineering is that as an epistemology, it is still to be organized and conformed. The best literature available is that to be found in the 1978 AAAS proceedings of the Macro-Engineering Symposium. Therefore, the first part of my talk this morning is to use three of last year's papers as a benchmark to establish the parameters for evaluating macro-engineering systems and then devote the remainder of this paper to selected topics and cases that I believe require development in order to mean big is beautiful as well as to provide means for developing the required epistemology.

Basic Evaluative Framework

Before proceeding with the main position of this paper, it is appropriate to set forth the background I have used in terms of the general state of modeling and analysis for macro-engineering. The general area of socio-economic policy systems modeling and analysis since 1970 is proving to be one of the major areas of academic research. There is much research work being devoted by many academicians to bring systems analysis and other methodologies to bear on a solution of macro- and micro-social problems. The impact of this work on common problems may well provide the tools and means for assessing whether either or both small and large macro-engineering is beautiful and how both large and small can benefit mankind.

Professor Kenneth E. Boulding succinctly stated the need and approach to the evaluation of large systems as follows:

> "The evaluation of large systems itself is a difficult problem, one which cannot easily be resolved by simple discussion or argument. Nevertheless, these remain questions of enormous importance, pregnant with potential for vast human misery, or perhaps great human satisfactions.

> "Improving our methods of evaluating these large systems, therefore, seems to be a high priority for the human race if we are to avoid what might be catastrophic error and ruinous decisions. At this stage I doubt very much if any general rules can be formulated to assist in these evaluations. Nevertheless, if the description of large systems can be improved, and if we can estimate the relative evaluative weights given different parts of the system, then we have at least some chance of breaking down the problem into smaller and perhaps more manageable pieces."[1]

It has been the primary assumption in this paper that the management of macro-engineering problems provides an adequate framework to describe macro-engineering systems that encompass their assessment in economic, social, political, and technological terms. In this respect, it is my opinion that macro-engineering has progressed from macro-engineering projects, to macro-engineering programs, and that our recent concerns with environment, resources, scarcity, social welfare, and national policies related to energy and technology are evidence of this project to program shift. My precept is that throughout time such needs and issues have had to be concerns of the senior executives and administrators of our major institutions.

---

[1] Kenneth E. Boulding, "The Evaluation of Large Systems," Technology Review, May 1975, p.68.

Let me now be more specific. Managers of
institutions exert authority through planning,
operations, and control systems. Those systems
simultaneously conform to and mold the manager's
conception of institutions, their purposes and
their needs.

A simplified model of a planning, operations
and control system can be shown as:

Institutional Planning and Control				
Resources	Planning	Operations	Control	Goods Services
	Desired Objective	Actual Transformation of Resources	Difference between plans and actuals	

In this model, inputs are defined as the
resources required to implement plans. These
resources are then transformed through operations
and controlled to meet the perceived objectives of
the institution.

Even in its detailed, real-world form the
foregoing model makes it hard for managers to perceive and analyze broad cultural, social and technical patterns in terms of their institutions'
desired objectives. Furthermore, the model's field
of view is too narrow to aid managers in seeing
their own and their institutions' roles in the context of society and its goals. In short, traditional conceptual models for management, though
useful in the past, are rapidly becoming limited
insofar as they are restricted to single objectives (primarily economic in nature), to self-
contained operational units, and to static environments.

In other words, the traditional models are
unable to provide management with the:

● means of conceptualizing and dealing with
  multiple objectives,

- means of conceptualizing and dealing with operations and interactions with other economic, social, cultural institutions,

- means of conceptualizing and operating within a highly dynamic environment.

Macro-engineering can give management the ability to overcome most of those restrictions. Some of the technology exists; some needs to be developed. The critical need is for constructs that will direct the advances and applications of macro-engineering so as to benefit management and the public.

It is the manner in which macro-engineering projects and programs have been managed in the past two decades that provides us with an adequate framework to evaluate macro-systems currently and for future needs.

Macro-Engineering Revisited

Three papers presented last year provide an excellent starting point to revisit macro-engineering. These papers are "Historical Perspectives on Macro-Engineering Projects" by E.S. Ferguson,[2] "The Planning and Conduct of Macro-Engineering Projects" by R.P. Godwin,[3] and "The Finance of Macro-Projects" by W.O. Sellers.[4] These papers,in some respects,established some of the more critical parameters in regard to the understanding and practices of macro-engineering projects.

Dr. Ferguson defined macro-engineering projects as follows:

"A macro-engineering project (MEP) in any age is one that strains current capabilities and resources. It is at the outer limits of current 'state of the art'; it is expensive; and because it is so different from run-of-the-mill projects, the MEP furnishes engineers and technicians with a particular challenge and fascination."

---

[2]University of Delaware and The Hagley Museum.
[3]Vice President and Director, Bechtel Inc.
[4]Vice President, Merrill, Lynch, Pierce, Fenner & Smith, Inc.

His definition concentrates macro-engineering on
large, unique projects that involve newer technology for the first time or are very expensive.
Thereby, he would include projects simply because
they were physically big. What is interesting to
me is that Dr. Ferguson's definition could include
a project such as the English Channel Tunnel as a
macro-engineering project in 1751 when Nicolas
Desmarets won the Academy of Amiens prize for the
test plan for a "dry shod" connection with the
British Isles, but not in 1979 when it is generally
accepted that the tunnel could be bored by very
conventional methods. It is for this reason in
revisiting macro-engineering that I find R.P.
Godwin's definition more useful to develop a
required epistemology and practices in that he
defines a macro-engineering project as "a large and
complex engineering project." Godwin's examples of
macro-engineering wonders are more closely allied
to Dr. Ferguson's general definition of macro-
engineering projects.

Godwin also distinguished between macro-
engineering projects and programs. His distinctions between projects and programs included similarities and differences. Their similarities are
that each is (1) complex; (2) large as to size or
scope or both; (3) takes a long time to implement,
generally measured in decades; (4) requires significant resources in terms of manpower, material,
and equipment, resources that can be measured in
the billions of dollars; (5) government involvement
either as client before projects can begin and/or
regulatory; (6) government involvement in financing
as his lower limit for macro-engineering projects
is $10 billion; and, finally (7) has substantial
social, cultural, economic and environmental
impacts.

The differences he sets forth between macro-
engineering projects and programs can be categor-
ized as to _first_ _scope_. Projects are generally
single-purpose facilities that are confined to a
specific site. Programs are national in scope
that are multiple-purpose projects that spread over
several sites for the purpose of developing hardware or some ultimate product. A second category
that distinguishes projects and programs is funding.
Projects are funded separately and from worldwide
financial institutions. Programs are funded from a

national treasury. Perhaps examples of macro-
engineering projects and programs can help to clar-
ify his distinctions. Examples of projects he used
are the Pyramid of Cheops, the Great Wall of China,
the Panama Canal, the Jubail Industrial Complex,
the Trans-Canada Pipeline and the James Bay Hydro-
electric Project. His macro-engineering program
examples are the Apollo Program, the Manhattan
Project, and the national highway program

In my opinion, it is implicit in Mr. Godwin's
differentiation between projects and programs that
there is more technology that can be diffused to
other societal uses through macro-engineering programs than through macro-engineering projects.
Projects employ more proven state-of-the-art technology while programs are concerned with both
cutting-edge technology, state-of-the-art technology and required advances of selected scientific
breakthroughs. The financing, management and
assessment of projects and programs can thereby be
differentiated. Therefore, in revisiting macro-
engineering, it is important that the epistemology
does distinguish between projects and programs.
Their similarities often have masked their significant characteristics.

The majority of last year's symposia was
concerned more with the macro-engineering projects
than with macro-engineering programs. In my
preparation of this paper on the evaluation of
macro-systems, it became necessary to review some
case studies of macro-engineering programs for purposes of contributing towards a body of knowledge
on macro-systems. For purposes of this paper, I
have placed more emphasis on my review of the
management, planning, and conduct of macro-
engineering programs.

Macro-Engineering Programs and Institutional Complexes

Macro-engineering programs generally originated from the need of a nation to meet a single
objective. More often than not, these needs arose
from crises that involved national defense or
national prestige and were given top national
priorities. Examples of such programs were the
Manhattan Project and the Apollo Program. These

programs had definite end goals which could be
perceived by the program managers as well as by
most segments of the national public. These macro-
programmatic national demands were expressed through
the medium of one aggregative consumer institution-
the Federal government. The government generally
established a new institution to plan the program,
conduct the program, and control the program including its financial, management, engineering and
quality controls. In response to the social
demands as exemplified by macro-engineering programs, a limited number of institutional suppliers,
in terms of system managers, R & D managers, equipment suppliers, and component manufacturers, et al.,
evolved in a relatively short period. By the time
President Eisenhower retired from office, he
referred to these individual supply-oriented institutions as clusters and the associated federal agencies as a complex; e.g., the military-industrial
complex. He went on further to call attention that
these complexes could well try to perpetuate themselves. The same general evaluative questions are
being asked today by various groups of other programs such as TVA, mass transit programs, welfare,
and other macro-programs.[5] In the past, macro-
engineering programs have institutionalized themselves both in terms of program management and in
terms of an institutional constituency comprised of
individual education and business entities that
view the program as a demand for their services
and goods. Institutionalization of macro-engineering programs through institutional complexes is an
emergent phenomenon since the 1960s.

One can model the various goals for such an
institutional complex in terms of macro-systems.
However, the limitations in models are encountered
in the situation where we must know when to change

---

[5]"TVA Today: Former Reformers in an Era of Expensive Electricity," *Science*, Vol.194, pp.814-818.
F.C. Colcord,Jr., "Institutions for Urban Transportation," *Technology Review*, Oct/Nov. 1973,
pp.58-59. A. Altshuler, "The Politics of Urban
Transportation Innovation," *Technology Review*, May
1977, pp.51-57. E. Herst, "Transportation Energy
Conservation Policies, "*Science*, Vol.192, pp.19-20.
R. Morris, "Welfare Reform 1973: The Social Services Dimension," *Science*, Vol. 181, pp. 515-522.

the goals that the macro-system represents. To
date, there are no models to handle these classes
of problems. Part of the reason for the lack of
such models is that the current models were based
on optimal planning solutions to societal problems
which these programs then represented. Because
macro-engineering problems take a long time for
solution the emphasis has been on successful solution of the problems of a previous period and not
on the impact that their solution created on present and future needs.

Willis W. Harman and his colleagues at Stanford Research Institute have identified, in their
analysis of social problems, the following four
dilemmas inherent in past macro-system solutions;
namely[6],

1) <u>The growth dilemma</u>. This dilemma centers
   on the need for continued economic growth
   but that our society cannot live with
   consequences; e.g., environmental and
   social costs.
2) The <u>control dilemma</u>. There is a need to
   guide technological innovation but we shun
   centralized control.
3) <u>The distribution dilemma</u>. There is no
   suitable mechanism or philosophy within
   the industrial system for redistribution
   between highly industrialized nations or
   regions with the less developed nations or
   regions.
4) <u>The work-roles dilemma</u>. Our society is
   increasingly unable to supply an adequate
   number of meaningful social roles.

These four dilemmas make up what Harman refers
to as the world macro-problem.

"Industrial societies in general, and
the U.S. in particular, are faced with one
fundamental problem which is so pervasive
and so pernicious that the related societal
problems (e.g., poverty, unemployment,
inflation, environmental deterioration,
crime, alienation) will defeat all attempts
at solution until it is satisfactorily

---

[6]W.W. Harman, "The Coming Transformation," <u>The
Futurist</u>, February 1977, pp.5-7.

resolved. The problem is not new, but
industrial and technological advances
have given it a new urgency.

"Individuals, corporations, government agencies and others in the course of
their activities make underlinedecisions
(e.g., to buy a certain product, to employ
a person for a particular task) which
combine to form macrodecisions of the
overall society (e.g., a 5% growth rate,
deteriorating cities, polluted air.)

"The world macroproblem now is that
perfectly reasonable microdecisions
currently are adding up to largely unsatis-
factory macrodecisions. The macroproblem
will be the predominant concern of the
foreseeable future for all alternative
paths. It is the composite of all the
problems which have been brought about
by a combination of large-scale industrial
development and high population levels.

"Although such terms as 'environment'
and 'technology assessment' have entered
the political rhetoric, we have not yet
begun to take the macroproblem seriously."[7]

Macro-Engineering and Social Engineering

For economy of time, let me generalize what I
believe is a next evolutionary step within macro-
engineering. In this regard, it is necessary to
explicitly recognize the implicit hierarchy underlying this discussion. This hierarchy is macro-
engineering projects, macro-engineering programs,
and macro-systems. The linkage among these three
levels of macro-structures is contained in the
management dimension. In my view, macro-engineering management evolved from macro-engineering projects management to macro-engineering program
management and is now evolving macro-systems
management. The implication here is that the principles of macro-engineering project management are
developed at the macro-engineering program level.
Analogously, the principles of macro-engineering
program management are to be developed at the
macro-systems level.

---

[7] *Ibid*, pp. 7-8.

The management of projects is well recorded in the literature and there are sufficient good and bad examples for study. To a lesser extent this is also true of macro-engineering programs. However, there is need to identify, to evaluate, and to extend our knowledge base with regard to macro-systems management. Macro-system management is to study the manner in which government, business, and educational institutions can be utilized to meet the demands of society that involve large-scale needs such as security, environment, education, and social,cultural and natural resources while maintaining a free, modern society.

Another way of making the same point is to say that micro-engineering must be co-joined with social engineering at a macro-system level so that newer institutional arrangements -- economic, political, social--can be formed that solve today's and tomorrow's human and social needs. This will require newer ways of establishing multi-goals for macro-systems. The goal setting is no longer simply in economic terms or social terms, but must establish political, technological and cultural goals.

An example of social engineering was provided by Jack C. Page, vice president of Booz, Allen and Hamilton, Inc.

"Air and automotive travel have inherent advantages over rail travel. The management and labor practices in the respective industries affect their performances. But when it takes longer to fly from New York to Philadelphia now than it did 30 years ago, and when our cities have monumental traffic jams, it is obvious we have not reached the optimum balance among these three modes.
"Suppose that gasoline tax revenues were used to develop better railroad terminals, reservations, signalling and communications equipment; railroad taxes to finance aircraft navigational aids and landing facilities; airline tax revenues to build new roads and super highways. What would happen?

"Under present conditions, the construction of new aircraft navigational
facilities and terminals would almost
cease while revenues would still flow
in to complete the highway building
program. Meanwhile, the railroads would
have substantial funds to improve until
they could compete. Their success would
again provide revenues to improve the
airlines. A stable transportation system
would ensue; and the relative emphasis
on different modes would be determined
by the relative tax rates. Raising the
gasoline tax would improve rail service
which would, in turn, improve air service,
and a new stable operating condition would
exist with decreased emphasis on the
automobile."[8]

Another reason for looking at macro-engineering from a macro-system and its institutional
complexes is that it can help develop required
national policies. J.E. Goldman, senior vice
president for R & D at Xerox Corporation stated
the following regarding development of a national
technology policy:

"As we look back over the past decade,
we have good reason to be uncomfortable
with our national investment in technology...
My own reasons for concern are twofold.
"First, it seems to me we have given
far too little support to the creation
of new technologies, including too little
toward the support of science itself.
Rather, we have spent our dollars--and
the energies of our technical people--on
the exploitation of given features of the
old technologies, principally through
scale-up and increasing reliability.
"Second, our technical priorities have
held relatively stable, despite the fact
that the world about us is undergoing
enormous change. During this past decade,
to cite one specific fact, our national
government has invested some 3 billion

---

[8]Jack C. Page, "Engineering Social Systems,"
Technology Review, July/August 1972, p.46.

to 4 billion man-years in research and
development (R & D) programs; less than
1 percent of this enormous investment
has gone toward R & D support relating
to such critical problems as housing,
crime, the urban environment, and ground
transportation.

"As we correct this imbalance and
begin now to channel a larger part of
our technical effort toward programs of
social significance, we must remember
one of the important lessons of our
recent past: the history of the space
program, it seems to me, is a lesson in
the mastery of the institutional techniques necessary to bring together the
segments of the intellectual, industrial,
and technological community needed to
fulfill goals in a timely fashion."[9]

Macro-Engineering Financing and Policy Initiatives

The financing of macro-systems is as complicated as the assessment of macro-engineering projects, programs and systems. Wallace O. Sellers,
in his 1978 macro-engineering symposia paper
entitled, "The Financing of Macro-Projects,"
pointed out that there are over 200 macro-projects
that were in the hard planning stages or under
consideration which will cost over half a billion
dollars each. Many today are over $10-25 billion
each. Energy program investments will easily
consume over 30% of all savings over the next
decade.[10] While Sellers indicated that financial
institutions were "no longer intimidated by large
numbers," he did set forth the factors which needed
to be considered before successful financing could
be achieved. Again, these factors require

---

[9] J.E. Goldman, "Toward a National Technology
Policy," Science, Vol. 177, p.1078.

[10] G. Kozmetsky and E.B. Konecci, "National Energy
Plan and Investment Analyses," Preliminary Assess-
ment of the President's National Energy Plan,
May 11, 1977, The University of Texas at Austin.

inter-institutional macro-system coordination, planning, construction, and control.

Today there is urgent need to consider and develop integrated macro-system management. The burden of responsibility and the authority to initiate the needed macro-systems rest squarely on our senior executives. <u>They not only have the</u> <u>formal ability to reallocate resources but are, in</u> <u>fact, the single key element required to guide and</u> <u>to assess the policies and programs and plans in a</u> <u>rational manner.</u>

The challenges of the macro-system programs will not be met or overcome by the use of polls, holding symposia, or conferences, conducting panels of expert predictions or in the use of casual "business as usual" staff work. Nor will they be solved by external assessment expressions such as holding protests, lengthy court battles, etc. The urgency of the problems themselves will no longer permit such luxury in the management of change. The methods must change. The requirement is for personal involvement at the levels of senior executives who are concerned with their own institutional policy formulation and decision making. Selected groups of executives from the major institutions must work in collaboration and in concert to formulate the programs and to take the first steps to resolve them. These working committees of executives are not one-shot affairs. They are continuing affairs. We know other nations have developed policy centers as working committees of senior executives; in particular, Japan. Japan has established the following policy centers:[11]

1. Public and Financial Subcommittee
   a. Working Group on Allocation of Investment
   b. Working Group on Social Overhead Capital

2. Industry Subcommittee
   a. Working Group on Information
   b. Working Group on Resources

[11]Taken from "New Economic and Social Development Plan 1970-1975," Economic Planning Agency, Government of Japan, pp. 163-185.

    c. Working Group on Progress of
       Technology
    d. Working Group on Labor Force
    e. Working Group on Small Business and
       Distribution
    f. Working Group on Agriculture

3. National Livelihood of People Subcommittee
    a. Working Group on Level of National
       Livelihood of People
    b. Working Group on Environmental
       Pollution
    c. Working Group on Urbanization
    d. Working Group on Social Security
    e. Working Group on Social Tension

4. International Economy Subcommittee
    a. Working Group on International Capital
       Movement
    b. Working Group on Economic Cooperation

5. Research Committee re:
    a. Fundamental Problems of Economic
       Planning
    b. Industrial location
    c. Land Policy
    d. Prices, Wages, Income and Productivity

Memberships in the above groups are made up of
senior executives from business, education, government, and other institutions. Moreover, they
relate to the problems and formulate policies to be
implemented by business, education, government and
other institutions with definite goals and milestones with time limits that provide for assessment of the effectiveness of their policies.

These barely visible emerging complexes are
creating the need and perception for broader and
more deliberately and skillfully designed goals and
criteria for the formulation of institutions and
their interrelationships. Current macro-engineering systems are no longer self-contained but are
highly interrelated. This situation would seem to
be tailor-made for a systems analysis; but it might
be noted that in order to carry out a systems analysis, you first need a system to analyze. And as
was previously indicated, neither policy centers
nor society's problems have been systematized.
While a partial solution includes the need for the

newer institutional complexes that are yet to be
clearly defined and structured, we must be aware
that many of today's societal problems are
recurring. This necessitates retaining many of
today's institutions and their interrelationships.
The newer institutional complexes are therefore,
an added layer over today's existing institutions
thus creating the need for an order of magnitude
of management abilities that have not been required
heretofore. What is more imposing is that such
management has had no development to aid them in
structuring the problems for solution other than
those which can be extended from macro-systems.
What this means for society is that:

1. Macro-systems present opportunities
   in the form of new products, new
   services and new markets for both the
   existing institutions and the emerging
   new institutions.

2. Government is both a generator and a
   consumer of the services and goods
   derived from the solutions of the
   skyrocketing demands (e.g., demands
   for minimum levels of housing, food,
   health care, education, highways,
   protection, etc.)

3. We can anticipate changes in public
   policy that provide means for attaining
   the desired partnership among government, industry, labor, and the public
   in strengthening the well-being of our
   nation.

Overview

The time is long past for the installation of common-sense methods for evaluating macro-projects and for the adoption of realistic guidelines to decision-making. According to William J. Jones, the evaluation process must consider the motivation for a large-scale project, its value to society as a whole, its practical risks, and the likelihood of changing or halting a macro-engineering effort once construction has been authorized. Government is often heavily involved, either through funding, regulation or guarantees, and the political process therefore plays a critical role. The evaluative process, if it is to be candid, must identify political forces supported by the special interests of various groups: the vocabulary of systems science should delineate--and not obscure--the underlying realities in a society where public decisions are responsive to constituency pressures.

3. Macro-Engineering: How to Decide

The epilogue of <u>Dividends from Space</u> describes a letter from a nun to a U.S. space scientist asking how one could justify spending billions of dollars in space research while millions of children were starving on earth.

In his reply, the scientist compared the space budget with the far greater expenditures by Americans for cosmetics and tobacco and stated that space activities should be less vulnerable in the scale of values than lipstick and the kingsize filter-tip cigarette, and that even if the space program were to be dismantled, social problems such as hunger, crime, urban decay, pollution, etc. would not go away.

Macro-engineering projects are too often proposed as heroic enterprises and excluded from adequate evaluation by stating that conventional yardsticks used to evaluate and cost projects, and rules of management and budget control, do not apply. Because of the category into which the proposals fall, they will always be for the benefit of all mankind and the "benefits" are needed!

It is the intent of this paper to suggest that the time is long past for development of methods for evaluating macro-engineering projects and the early adoption of some guidelines in decision-making, even though they may be rudimentary and will require considerable improvement.

Macro (meaning large or extensive), as applied to engineering, can describe size, technical difficulty, time required for design and construction,

initial costs in money, size of labor force, time
to completion, magnitude of impact, etc.

Some projects are revealed as macro-engineering efforts only when viewed after completion. New
York City, London, Rome, etc. fall into this class;
they "grew" without a prior prepared detailed blue-
print. Washington, D.C., Brasilia, and the proposed new capital of Nigeria, are examples of progressively greater degrees of city pre-planning.

"Hot" and "cold" wars involve macro-engineering. Such efforts are characterized as being based
almost exclusively on the perceived comparative
end-effectiveness of projects. Costs in dollars,
environmental impact, etc., are relatively minor,
if at all, considerations.

Appendices A,B,C, and D describe other macro-
engineering projects. To fall under this classification, a project need not be large in itself. It
can involve a very small effort and have a very
large impact. "The Pill", incandescent electric
light lamp, internal combustion engine, "Coca-Cola"
are examples.

Macro-engineering includes all projects which
have sizeable impacts on society, the economy, the
environment, governments and lifestyles.

The classification of impacts into first
order, second order, and higher order groups changes with time, political winds, and the interaction with other such projects and frequently
singularly insignificant events, discoveries or
ignorances.

A Definition of Macro-Engineering

I suggest that a project can be defined as
"macro-engineering" if it falls under one of the
following categories:

1) It entails government funding, or other government involvement (guarantees, special tax incentives, etc.) because of the magnitude of capital
investment requirements, the extent of environmental impact, the time span to completion.

2) Uses exhaustible natural resources that may be

renewable but only after a long gestation period.

3)   Large numbers of the population or particular
segments of society will be affected.

4)   Requires substantial participation of state
governments, foreign governments.

5)   Obligates the government to monitor, control,
safeguard the products, plant or residue for long
periods of time or in the event of failure of the
private sector to do so.

Why Should We Be Concerned?

It is this author's suspicion that we have
reached, or have even overstepped, the limits of
our economic, manpower-management, and social
tolerance limits.

There is pre-occupation with government funding of politico-economic innovation for achieving
national goals: war, energy supply, transportation,
space exploration, social welfare (health, recreation, etc.).

There is increasing national and international
pressure for spectacular technical advances.

In the next 15 years we must mobilize as many
raw materials as have been extracted during all of
man's previous history on this planet.

Within the next 10-15 years we must design,
manufacture, install and bring into full operation
as much power production equipment as has been
accumulated up to this point in our history.

A characteristic of technological advancement
is a decreasing requirement for labor in production, more sophisticated scientifically-intricate
production means and increasing attention of scientists to research and development. Every LDC wants
to make a quantum leap into modern industrialized-
nation status. The rising expectations of the
masses of peoples of the Third World are encouraged
and accelerated by television, movies, and newsprint. [The standards of living in established
countries of the First and Second Worlds are
threatened by shortages in natural resources and

controls of supply in countries of the Third World.]

The commitment of capital to construct and tax support to operate, regulate, monitor and dismantle, strains the national economy, the social fabric, political security and the physical and biological structure of our planet and its atmosphere.

There is a centralization of decision making by persons whose accountability in time is much shorter than the time required for demonstration of failure or success; injudiciousness of the original decision becomes apparent only when the "ins" are "outs."

Motivation

The motivations for initiating macro-engineering projects may be classed under personal desire for power, conceit, religion, monetary profit, political stability or advantage, national pride, competition, growth, health, safety, etc., including "the good of mankind."

For a number of nations there is no demarcation between theological and political concerns-- they are one. With others, political philosophy and action have replaced conventional concepts of religion without the population's being fully aware of the event. In many cases, engineering projects serve only to advance the established "religion" or provide more support for the prevailing "theology/politics".

The most powerful motives are of a competitive sort: "We want to come out on top." We want to be able to credit ourselves with first place, to be second is bad, and to be second-rate is intolerable. Motives such as the promotion of scientific, technological and economic progress are less compelling in political circles, though in other circles one or another of them may be dominant.

The competitive spirit and the desire to excel are important and an integral part of American life. There has always been an element of emotional commitment to big projects. If a project is seen as a challenge, the view is that it is good to excel, that it is good to test one's mettle against significant challenges.

Political Stabilization, Gains and Diversion

Large projects are apt to evoke domestic enthusiasm and national pride and awe and respect from the leaders of many of the nations of the world. They can be counted upon to gain international prestige and augment national pride with all shades of opinion, reflecting differing philosophies and sets of values across a broad spectrum. At one extreme, the development of a project is a technical problem resolvable through engineering and economic efficiency. At the other extreme, symbols for international cooperation and greater socio-political understanding and progress appear to dominate the evolution of a "high technology" program.

That leaders and political parties seek to promote their own political fortunes need not surprise us. The "ins" champion policies (goals and rules and methods for achieving them) with an eye to the next election or crisis. The "outs" look for weaknesses, failings and omissions. They try to devise alternative policies with public appeal.

Political concern does not focus primarily on the scientific and technological measures to implement a program often asserted in generalities.

We should be more curious about the values and interests -- the motives or reasons -- that inspire political behavior, especially in the field of international relations. To what extent is a project motivated by a desire to enhance American prestige over the world or to have one's name recorded in the annals of history?

The technical characteristics of the military arms industry have changed since 1955:

Salaried employees	25% to 46%
Engineers and Scientists	10% to 16%
Floorspace devoted to manufacturing	52% to 40%
Laboratories and Offices	18% to 28%
Manager, schedule controller, procurement and overhead services people	14% to 29%

The changes indicated above imply a parallel shift in the composition of many legislative constituencies. The political implications are obvious.

We must try to evaluate giant projects associated with modern warfare in terms other than profit and employment opportunities.

By enlisting private organizations in the performance of public functions, government involves them in politics and blurs the line between "public" and "private" industry. The private organizations, however, retain the advantages of private enterprise while serving the vital needs of the nation and still influencing their own futures. How do we deal with this "grey area?"

Is it physically, politically and economically possible to build, operate, and maintain future macro-engineering projects in a common-sense manner?

The dilemma of the situation in which we find ourselves was expressed by a person most involved in the space program:

"We need to have and understand nationally-accepted goals or purposes.

How can we decide how important it is to spend, on an urgent basis, the very large sums of money required to put a man into orbit, etc., unless we have a pretty firm grasp of what the purpose behind the whole space effort really is?

And yet, who knows the answers to this and many similar questions today? Who is thinking about them and doing something about developing some answers?"[1]

When a project is advocated, how do we determine:
a) that it will accomplish what its promoters claim;
b) what hidden advantages are there, i.e., advantages other than those suggested by the proposers;
c) what common yardstick of "values" can be used to compare one project with others;

---

[1]T. Keith Glennan, First NASA Administrator.

d)   what are the "true values" of each of the
several classes of costs.

Value

What is the value of a project? How do the
values of one project compare with another's? How
can one compare different values? How can one
apply a number to a value? What is a value "worth"
in dollars?

There is no absolute way of proving or dis-
proving one or another goal value. Some have only
temporary or future values. One project has values
only when compared with the values of another project. Some values exist only at the sacrifice of
others. To propose projects for security, prestige
and pride is patriotic; to challenge projects that
are purported to be for those ends is unpatriotic.

1)   Technology is becoming more voluminous and more
complicated.
2)   Large-scale engineering projects have completion dates too far into the future to permit adequate assessment or no timetable at all.
3)   Much new technology is so complex and the
time span to stages which permit reasonable evaluation is so wide or indefinite that it is extremely
difficult to anticipate how it will do its primary
job and what its second-order consequences will be.
4)   As our understanding of biological, ecological,
economic and social processes improves, as we observe and realize the immediate consequences, we
have an obligation, under our planet stewardship
responsibilities, to evaluate, to the best of our
abilities, the second-order consequences of all
our actions, and to include their costs, monetary
and otherwise, in our analysis and decision
process.

Exploration of space has been compared with
the voyages of Columbus who was funded by Queen
Isabella who, in turn, was funded by peasants.

There were many other explorations which we
do not hear about. Expeditions that were very
expensive to initiate and ended in disaster.

Proponents of a project will always compare it
with a highly successful one. The evaluation

should compare it equally with those that were unsuccessful.

Many of the programs being advocated at this moment (for example, the world-wide conversion to coal as an energy source) have planetary phenomena intrinsically associated with them. Since the impacts will be global, control and management are likely to be effective only if they can be organized on a world-wide scale. Full achievement now or within a foreseeable future is not possible; nevertheless, even partial consensus is desirable.

Most of the major public engineering expenditure decisions have been characterized by "muddle through" and "rule of thumb". Public money has been lavished often on "popular" projects with a very hazy idea of the return to be expected and even the extent of all construction costs and obligations once completed. Those who object are stigmatized as "wild-eyed idealists" or "vested interests."

There is continuing need to discover and explain the costs and benefits and to see that our social control system helps resources flow into the appropriate channels.

Dorfman warns, cost-benefit analysis may help but the art can be likened to the problem of appraising the quality of a horse-and-rabbit stew, the rabbit being those consequences that could be measured and evaluated numerically, and the horse "the amalgam of external effects, social, emotional and psychological impacts and historical and aesthetic considerations that be adjudged only roughly and subjectively." The horse is bound to dominate the flavor of the stew; meticulous evaluation of the rabbit is hardly worthwhile.

There is a major problem in evaluating macro-engineering projects (MEP). That is, the quantification of "externalities" and "group wants".

Non-compensated external costs must be addressed. "Group wants" either must be accepted as is or made to be founded on accurate and complete technical information.

There are inherent hazards in leaving of

decisions about MEP's in the corridors of political power. The establishment of economic and social priorities by Congress reflects primarily political pressures by vested interests. If one does have faith in the conventional political process, can we create tools which the Congress and government officials might be obligated to employ to maximize objective evaluation and action?

Is there a social rate of discount to be applied to MEP's because the magnitude of the investment is a sizeable portion of the GNP and the completion date of the project falls outside of the period of responsibility and accountability of the proponents? Can the MEP be analyzed within or in isolation from the rest of the economy? Can a decision to continue or discontinue an MEP be made in light of anticipated conditions at any point after it has begun?

The summary question is -- how does one determine the cost and benefit streams of government-sponsored macro-engineering projects?

The determination of rates of return on capital acts as one of the prime movers in resource allocation in the private sector of the economy. The government must develop a procedure for testing the "soundness" of macro-engineering projects, "soundness" being the desires of the public after being informed of all aspects.

Risk

A fundamental element of social control is the necessity for full and candid disclosure and discussion of potential costs and benefits. One cannot expect that those who are sponsoring the development of an MEP will indulge in listing its undesirable social consequences since an inherent feature of the promotion process is to minimize these consequences and to argue that they can be technologically overcome.

Investment by the private sector considers risk, the project will not "pay-back" as expected.

To what extent do the government-contract funded projects reflect consideration of risk? Are the proposers subject to penalty if the pay-back is not as promised?

With an MEP, where performance can only be measured many years after the start of the project and tens of millions, or in some cases, billions, of dollars have been invested, there is no "inordinate" risk perceived by the proponents.

Conclusion

Improving the process by which any sizeable federal program is conceived, evaluated, approved, developed, financed, managed and terminated is a challenging task.

The complex nature of our political system, the stubborn resistance to change inherent in a giant governmental bureaucracy, pose formidable obstacles.

With MEP's, the attendant cost of hundreds of millions and billions of dollars, and the time from conception to realization lead one to wonder whether anything at all can be done.

Nevertheless, the magnitude of the irreversible commitment to macro-projects which will steer policies for a few generations, and our involvement in the internal detailed affairs of other countries and the relationships between them and us, make it essential that we allocate a considerable portion of our attention and efforts to the task.

We must be concerned not only with efficiency (adherence to budgets and completion dates) but also to value (value related to all other economic and social needs and desires) and the objectives (sub-goals) as related to reaching long-term mission or direction for public policy.

The goals of public policy must be developed, specified and ratified by the political process as an expression of the people's will.

In order for that will to result in effective action, government officials and the public at large must participate in a continuing process of self-education.

Objectives and even goals must be revised as

more is learned about their costs and interrelationships.

There must be an awareness of and comparison
with alternatives to a proposed project. Valid
analysis requires fundamental research and experimentation on relationships between means and ends,
results and costs.

MEP's have output or cost implications that
extend significantly beyond the federal government's one-year operating and budgeting period,
more often than not, beyond the tenure of political
personnel and frequently into future generations.

In the execution of the program, responsibility must be assigned at all stages for costs, and
adherence to completion schedule.

Effective program progress evaluation requires standards of measurement, proper relative
weighting of the several criteria and data.

A comprehensive technology assessment must
precede the proposal of a project. Technology
assessment must be conducted at frequent "milestones" during construction and throughout its use
so that modification or termination is instituted
at the proper time.

a) Imbalances that arise during and after MEP's
are correctible only with great difficulty ex post
facto actions. Because of the magnitude of any or
all elements, risks are prone to become irreversible.
b) If it is determined or suspected that an MEP
endangers economic growth, environmental stability,
social quality of life, or turns out to constitute
a major menace of depletion of natural resources,
will inertia prevent modification of unpleasant-
nesses in time to avoid severe exacerbation?
c) Socially responsible management of an MEP is
virtually impossible. Too many facets of society
are affected, some positively, some negatively, in
major ways. More important, the Heisenberg uncertainty principle -- the precept that the accurate
measurement of an observable quantity necessarily
produces uncertainties in one's knowledge of the
values of other observables -- applies in social
behavior.

Government must learn to distinguish between:

1) regulations which protect the public's financial and other interests, and;
2) regulations which result in a loss of industrial incentive, creative ability and responsibility.

It must improve its relationship with industry by increased emphasis on competitive award of contracts and cost-plus incentive-fee contracts.

It must increase pressure for industry to assume a greater share of risk.

Government must pay increased attention to relatively unsophisticated components and systems which are in small sizes, impacts and dollar value. An outline of a partial test or evaluation of an MEP follows:

1) Need or desire: who
   a) wants
   b) advocates
   c) profits
   d) loses

2) Has the general public been given a chance to hear open debate, and an opportunity to register approval or disapproval?
   a) Alternatives. What can be done to accomplish same <u>end</u> result?

3) How much of "need" or "desire" is based on:
   a) transient conditions
   b) "soothsayers" prediction vision of the future
   c) creations by vested interests.

4) Who finances? and what are the risks during
   a) planning
   b) construction
   c) operation
   d) dismantling
      -- because of failure
      -- outliving usefulness

5) Is cancellation of project possible and where along its life do the possibilities exist?

There is a technological race among the powerful or would-be powerful, and an increasing gap between the "haves" and "have-nots."

There is a race between what might be called the technological factors of equalization and the countless disparities and inequalities -- military, economic, political, ethnic, ideological -- which distort this revaluation and often put it in the service of the most powerful and of the most dissatisfied.

Not all MEP's nor all aspects of a particular project can be evaluated in dispassionate, objective terms. There are casual conditions that are beyond human control, but the conscious reasons that people entertain concerning the wisdom of this or that action should be determined and understood.

Evaluation requires the constant presence of awareness that the programs are always, some to a greater degree than others, self-serving. In our society, we encourage competition and achievement by offering personal incentives (money, fame, power, etc.).

The programs involve tens of years so that the proponents' identities are lost. There is absolutely no liability for failure, no one and no organization is accountable. The original promises are forgotten.

The scientist stands to get research funds, the university anticipates grants, non-profit research organizations want contracts out of which they can pay high salaries, the trade union wants to keep up employment, the business concern wants profits, the trade-journal caters to the complex of readership, the congressman seeks re-election, the promoter (governor, elected official etc.) wants to bring more wealth to a state or region, and the U.S. President wants immortal fame -- beginning within his tenure in office.

Overall there is a very definite limit to growth, and within that overall limit, a limit to rate of growth. World demand is not only for energy, but for food, forest products, minerals, fresh water, marine protein, skilled labor and so forth,

as a function of rising expectations, rising afflu-
ence and rising population numbers. The technologies that underlie our economic system evolved in
a situation of relative resource abundance. The
real dilemma that we face is the need to impose a
rational and conscious choice upon the limited
alternatives which confront us.

"For which of you, intending to build a tower,
sitteth not down first, and counteth the cost,
whether there be sufficient to finish it?"
                                    (Luke 14: 28)

Appendix A
Trans-Planetary Subway Systems

Robert Salter of the Rand Corporation suggested, at last year's annual meeting of the AAAS, the adoption of the "Planetran" concept, a subway system moving at thousands of miles per hour, capable of crossing the United States in an hour or so. It was proposed as a possible alternative to the anticipated over-filled and hazardous airways/airports of the year 2078.

Who but the government could possibly begin to even explore the feasibility of such a project? Once commenced, who would be powerful and secure enough to stop funding if it seemed not to be a viable solution to the problems? Once the feasibility study of this concept is begun, what chance would there be for any other project to replace it even if the alternative were more promising? How close to the actual costs can estimates be? What could the government do if it found that costs were exceeding estimates?

How do we compare it with other proposed solutions? Not only technical characteristics but human preferences, prejudices and expected life styles which could change several times over the period required for project completion will have to be taken into account.

Appendix B
Erie Canal*

The Erie Canal was the work of that remarkable generation in America which made the period between 1815 and 1860 an age of great national expansion. It was a bold scheme designed to bring the Mississippi through a northern waterway, on American soil, to the growing metropolis at the mouth of the Hudson River. The supporters of the idea stated that the result would be national growth, strength and prosperity.

Travel between the west and the northeast coast cities was by poor roads, or in part, along the St. Lawrence River, which was closed by ice for a number of months and involved contact with the Canadian government and terminated in Canadian cities.

It cost $2.00 to send a barrel of flour 130 miles overland and the same barrel could go by water from Albany to New York City for 25 cents, a distance of 160 miles. A waterway between Buffalo and Albany would provide cheap, safe, "American-owned" transportation between the West and the East. A cannon worth $400 in Washington cost $2000 to transport to Lake Erie. At the Niagara frontier there was a foreign power, Canada, which controlled the only outlet of the Great Lakes to the ocean.

The Commission charged with evaluating the "worth" reported favorably and included the statement.... "After a lapse of two thousand years and the ravages of repeated revolutions...this national work shall remain...."

The report theme was: the potential productivity of the uninhabited Western lands, the pressing need for communication, the visions of private and public gain, the fear of Canadian rivalry. The project was funded by New York State and administered by a board of Commissioners, all with political ambitions.

---

* See the extended discussion of this project by Jeanne Krause in Chapter 7 of this volume.

It was estimated that the canal could be completed in ten or fifteen years at a cost of six million dollars. The project, with its thousands of jobs, was tailor-made to augment the power of the patronage in New York. It cost a little over seven million dollars, or about 16% more than estimated.

The canal did provide the promised communication with the West. Western products were exported through the canal to New York City in amounts greatly in excess of the original estimates. Merchandise reached the mid-west and the northwest from the East Coast through the canal. Emigrants traveled the route by the thousands and contributed to the rapid development of the West.

With the proliferation of railroads, the worth of the canal began to decline. It is non-existent today. But it exceeded its predicted contribution to the development of the West. The economist, W.W. Rostow, cited it as the principal contributor to the "take-off" stage of national economic growth in the 1840's.

Appendix C
Military Macro-Engineering

U.S. Military MEP efforts during both war and peace have been described as cooperative arrangements between the military, the arms industry and--frequently--political parties. It has been argued that military projects must be viewed with great suspicion and concern.

The proponents are described as having well-established vested interests and join in opposition to any form of disarmament plan, promote new and more expensive projects and sometimes appear to be engaged in an orchestrated "see-saw" escalation of an arms race with perceived enemy countries.

If there is any basis in fact for these charges, one wonders if it is at all possible to evaluate projects in an atmosphere devoid of profit motive, opportunities for the employment of scientists and engineers, much less the eternal issues of personal pride, ambition and the egos of political leaders.

One is tempted to ask whether France's spectacular commitment to the Maginot Line could not have been usefully replaced by the building of the armoured formations sought by iconoclasts such as (Colonel) De Gaulle!

Appendix D
"Lessons Learned from Constructing
the Trans-Alaska Oil Pipeline"

In a 1968 feasibility cost study a private
group estimated that an oil pipeline system from
Prudhoe Bay to Valdez, Alaska, would cost $1.046
billion for a 1.2 million-barrel-a-day capacity.
On the basis of this estimate decision makers
calculated the cost of a barrel of oil delivered
to the "lower 48", and concluded that the benefits
(including environmental impacts) were positive
and that it would be in the best interests of
the public to grant rights-of-way, through
federal and state-owned land, to a private company
(Alyeska Pipeline Service Company) to design and
construct the pipeline system.

Shortly after pipeline construction began in
1975, the company established a base control budget
of about $6.4 billion.

By December, 1977, at the completion of the
line, the cost was $7.9 billion.

It is argued that this project was privately
financed. The federal and state governments, on
the basis of cost-benefit analysis using the
original estimate, granted rights-of-way from the
Arctic to the Pacific Ocean.

The General Accounting Office, at the request
of the Senate Committee on Energy and Natural
Resources, reviewed the project and recommended
that the following should apply to similar future
projects:

a) First and subsequent cost estimates should be
    viewed with skepticism.
b) As much site-specific data as is economically
    practicable should be obtained.
c) Technical and geological uncertainties should
    be thoroughly investigated.
d) Government approval should be contingent on
    detailed planning for management control,
    including budgetary controls.
e) The Alaska Natural gas pipeline project's expen-
    ditures should have an ongoing government audit
    to protect the public interest.

Appendix E
Motivations for Adoption of a Project

Motivations for adoption of a project can be classed as follows:

a)   Military Security:

    1)   immediate: communication, navigation, transportation, and/or other basic capabilities
    2)   potential: new technology, weapons, etc.

b)   Peace

    Mitigation of military irritants or increased supply of shared geography, resources or opportunities

c)   Progress in Science and Technology

    1)   The desire for knowledge
    2)   Technological progress (technical efficiency)

d)   Economic and Social Progress

    1)   Direct: new products for home, industry (created desires, created markets)
    2)   Spill-over: new industries
    3)   Social Effects: health and safety

Miscellaneous Possibilities (Socio-Economic Mobility and Fantasies)

e)   National Prestige

    What, however, is the meaning and importance of "prestige"?
    In whose minds and to what ends?
    For what?

f)   National Pride

    1) achievement
    2) political (party) security
    3) challenge from others (imagined or real)

g) Special Interest and Ulterior Motives
   (Private and expressed)

   1) Personal incentives
   2) Private advantage
   3) Catering to constituencies
   4) Keeping going concerns going
   5) Serving partisan interests

Bibliography

"Dividends from Space"
F.T. Ordway, C.C. Adams, M.R. Sharpe
Thomas Crowell Co., N.Y. 1971
A "documentation" of the benefits of the Space
   Program

The Industry-Government Aerospace Relationship
SRI Project No. IS-4216, 1963.
Stanford Research Institute, Menlo Park, CA

The Politics and Technology of Satellite
   Communications
Jonathan F. Galloway
D.C. Heath & Co., Lexington, MA 1972

"A Technology Assessment Primer"
L. Kirchmayer, H. Linstone, W. Morsch, editors
No. 75 HJ 3137-7TFA Inst. of Elec. and Electronic
Engineers, N.Y., NY. 10017

"Measuring Benefits of Government Investments"
R. Dorfman (ed.)
Brookings Institution, Washington, DC 1965

"Lessons Learned from Constructing the Trans-
   Alaska Oil Pipeline"
Report of the Controller General to the Congress
EMD-78-52,
U.S. GAO, Washington , DC 6/15/78

Overview

Macro-engineering programs and projects pose difficult technical and managerial challenges at every stage of their development. C. Lawrence Meador and Arthur C. Parthé, Jr., argue that these challenges derive from several characteristics of macro-engineering projects and programs that are not found (or are only found in relatively minor ways) in conventional projects and programs. Macro-engineering projects often require significant quantities of public and private sector resources that must be committed (and thus put at risk) over relatively long time frames. Difficult technical and scientific problems must frequently be solved before the projects can be accomplished. This may result from the raw size of structures to be put in place, from systems complexity or from the emphasis on developing previously unknown technologies. Finally, macro-engineering programs may, and in fact often do, have profound socio-cultural impacts in the societies which develop them. This raises a host of issues including environmental implications, legal and regulatory factors, economic and political effects, to name only a few. Unfortunately, existing managerial systems, methodologies and frameworks do not address this portfolio of multi-dimensional objectives and constraints in an effective manner. A framework is presented which characterizes macro-engineering programs and projects from a managerial policy, planning and control perspective. From this framework a multi-dimensional and relativistic project state space is developed in which to portray analytic representations of macro-projects.

4. Managing Macro-Development: Policy, Planning and Control System Implications

Introduction

Macro-engineering programs and projects pose difficult technical and managerial challenges at every stage of development including planning, research, analysis, design, implementation, testing, evaluation, demonstration, and operational deployment. These challenges derive from several characteristics of macro-engineering projects and programs that are not found (or are only found in relatively minor ways) in conventional projects and programs. Macro-engineering efforts require that huge quantities of public and private sector resources be committed (and thus put at risk) over relatively long time frames.

Macro-engineering projects often require the solution of difficult technical and scientific problems before they can be accomplished. This may result from the raw size of structures to be put in place, from systems complexity of inter-acting components and subcomponents, or from the emphasis on developing previously unknown technologies.

Finally, macro-engineering programs may, and in fact often do, have profound socio-cultural impacts in the societies which develop them. This raises a host of issues including environmental implications, legal and regulatory factors, economic and political effects, to name only a few.

Unfortunately, existing managerial systems, methodologies, and frameworks do not address the portfolio of multidimensional objectives and

constraints which we have alluded to in an effective manner. At this stage, we need a framework
and a methodology to:

- Facilitate better policy analysis, planning,
  and control of macro-engineering projects
  and programs,

- Address non-financial and non-technical
  issues (e.g. legal, political, environmental,
  etc.), and

- Create an audit trail for the examination
  and resolution of problems and decisions
  arising during development of projects.

Further, we need to provide project and program
managers at all levels with better tools and techniques for effective technical and managerial
planning and control.

Decision Support Systems

Over the past several years there has been a
dramatic trend toward providing managers and technologists in complex problem and decision environments with computer-based systems which help to
organize, store, process and present information
relevant to planning and control issues which they
face. Recently, a new trend has emerged which
focuses on difficult semi-structured problem
environments where programmable solutions are not
available which "solve" the problems addressed. In
some of these cases a computer aid can be used in
various modes to support the problem-solving and
decision-making activity that is required. Scott
Morton calls this new trend in the development of
computer-based decision aids the Decision Support
System approach.[1] We believe that decision
support systems have exciting potential for addressing many of the difficult macro-engineering planning and control issues which we allude to above.

The balance of this discussion presents the
rudiments of a framework for managerial planning

---

[1]See, for example, Keen, P.G.W. and M.S. Scott
Morton, Decision Support Systems, Addison-Wesley,
Reading, MA, 1978.

and control of technology development in macro-
engineering projects. The major dimensions of the
framework include issues related to project cultural impact potential, technology development
difficulty, and public/private resource requirements. This framework is implemented in a new
emerging project planning and control decision
support system called MACRO-PLANNER which is
currently under study and development at MIT and
the Charles Stark Draper Laboratory, Inc.

The Framework

Figure 1 presents a three-dimensional visual-
ization of what we shall call macro-engineering
state space. It can be thought of as an abstraction of the key dimensions through which macro-
engineering projects or programs must be evaluated.
We can further refine these dimensions by listing
their key components in the following ways:

Cultural Impact Potential
        Environmental Implications
        Political Effects
        Life Style Impacts
        Legal and Regulatory Results
        Changes in Attitudes, Perceptions,
          Options and Opinions

Technology Development Difficulty
        Fundamental Science Barriers
        Systems Complexity-Size/Structure
        Interdependency Problems
        Modelability - Physical and Analytical
        Empirical versus Theoretical Emphasis

Public/Private Resource Requirements
        Manpower - Technical, Managerial and Other
        Funding from Public and Private Communities
        Materials, Supplies, Critical Raw and
          Manufactured Goods
        Governmental Inspection and Regulation
          Requirements
        Energy Demands

A macro-engineering project can be thought of
as "extended" in macroengineering state space as a
function of the magnitude or the degree to which
individual components of the state space are
impacted. Figure 2 suggests how three macro-

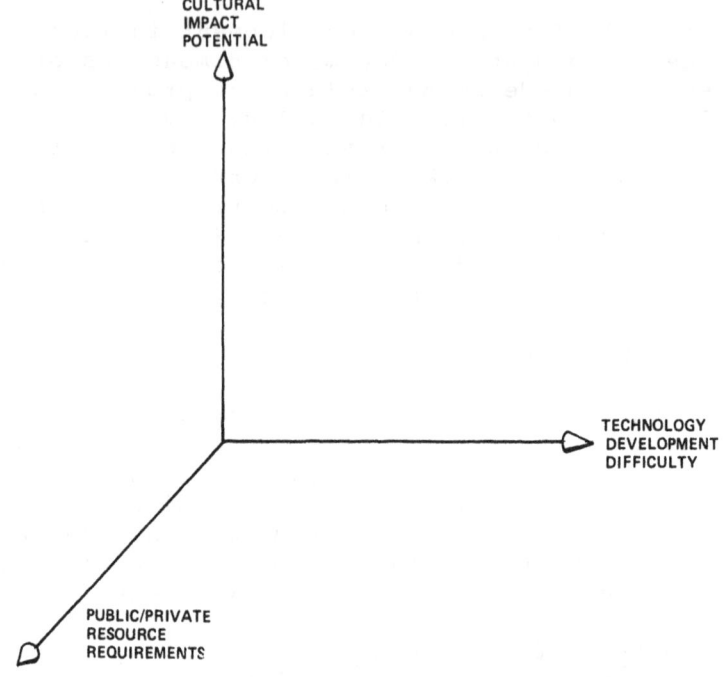

Figure 1.   Graphical representation of macro-engineering state space.

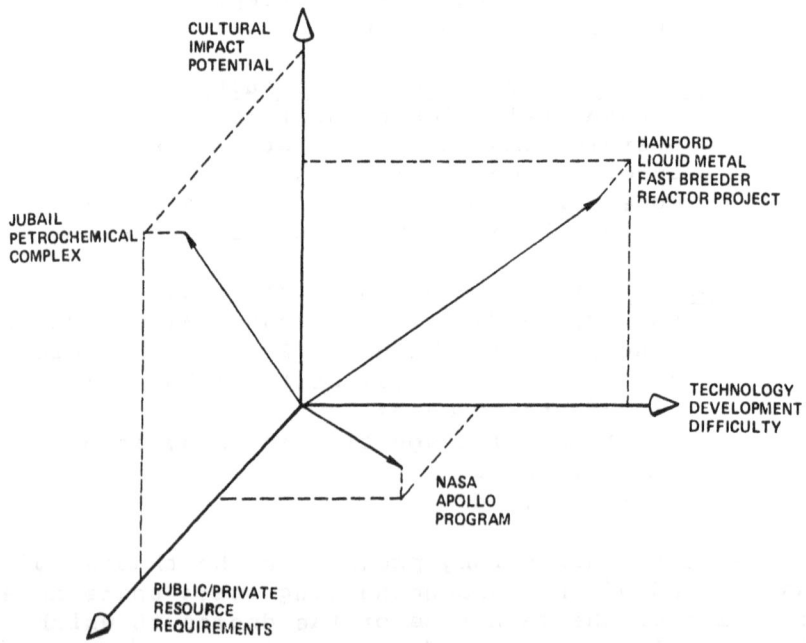

Figure 2.   Comparison of macro-engineering efforts.

engineering projects might be viewed along these
dimensions in comparison to each other. The dimensions of cultural impact potential, technology
development difficulty, and public/private resource
requirements are only relevant when considered in
the relativistic context of the socio-cultural
milieu in which they are evaluated. The building
of the Pyramid to the Sun in the Valley of Mexico
by the Teotihuacan or the construction of the
astronomical observatory in Chichen Itza on the
Yucatan peninsula by the Maya must certainly be
viewed as macro-engineering efforts at their time
and place of deployment. The Teotihuacan pyramid
was equivalent in relative socio-cultural impact
and resource requirements to the present day Saudi
Jubail petrochemicals complex. The Mayan astronomical observatory required resources and technological or scientific accomplishments equivalent
in relativistic terms perhaps to the U.S. Apollo
program.

The Information Hierarchy

   To be useful for developing meaningful and
relevant insights and guidance toward planning and
controlling the deployment of macro-engineering
programs and projects, a framework must be sensitive to the type, quantity, and characteristics
of information appropriate to different parts of a
macro-engineering organization. Figure 3 illustrates the information hierarchy pyramid which
characterizes the policy and strategic planning,
program management control, and project operational
control functions necessary in a macro-engineering
organization. The implication of the pyramid is
that a great deal of data (and people and organization units) exists at the lowest operational
levels but that this information is only necessary
and appropriate in highly aggregate form for
decisions of a strategic or policy type at the top.

   Figure 4 demonstrates the very important principle that different managers within a macro-engineering organization may have dramatically different information needs depending on their level and
type of responsibility, organizational location,
and personal managerial "style." In attempting to
develop a macro-engineering planning and control
decision support system, it is essential that such
differences be taken into account. The implementa-

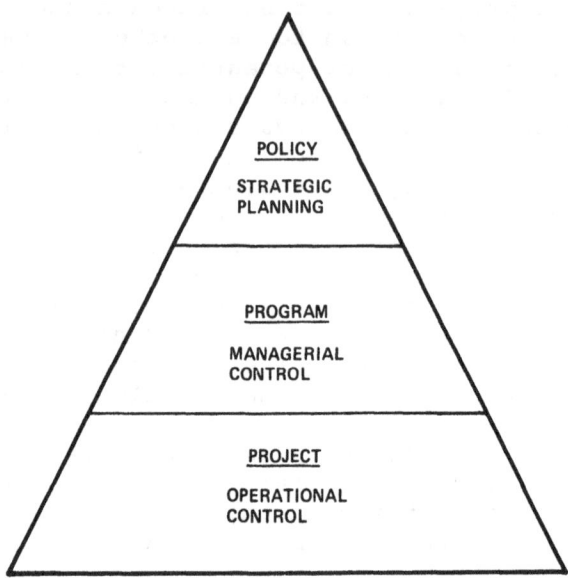

Figure 3.   Information hierarchy.

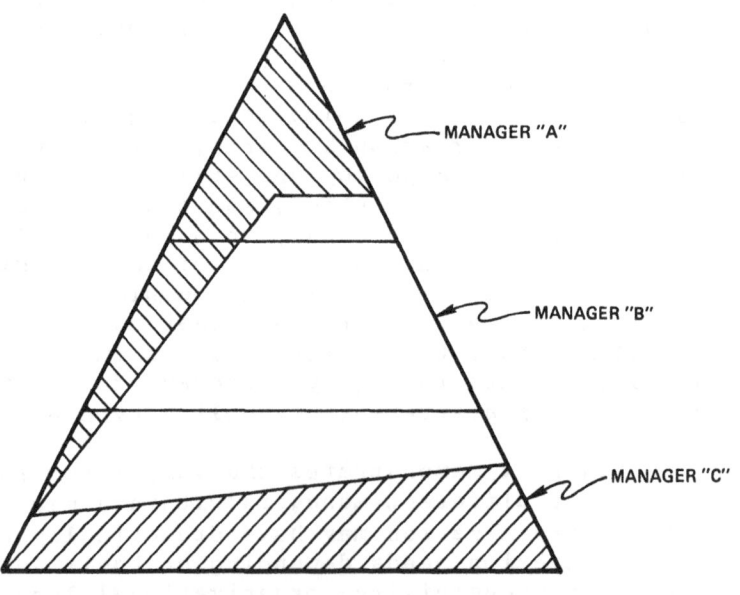

Figure 4.   Differing managerial
information requirements.

tion of multiple models and information processing
options for addressing problems from different perspectives is a useful and relevant way of accom-
plishing this objective in a decision support
system environment. More will be said on this
issue as we describe our own efforts in developing
the MACRO-PLANNER system described later.

Structured Versus Unstructured Decision Problems

An important though somewhat hazy distinction
can be made between so-called programmed and nonprogrammed decisions:[2]

> Decisions are programmed to the extent
> that they are repetitive and routine,
> to the extent that a definite procedure
> has been worked out for handling them
> so that they don't have to be treated
> "de novo" each time they occur...
> Decisions are nonprogrammed to the extent
> that they are novel, unstructured, and
> consequential. There is no cut-and-
> tried method of handling the problem be-
> cause it hasn't arisen before, or because
> its precise nature and structure are elu-
> sive and complex, or because it is so
> important that it deserves a custom-
> tailored treatment. By nonprogrammed I
> mean a response where the system has no
> specific procedure to deal with situations
> like the one at hand, but must fall back
> on whatever general capacity it has for
> intelligent, adaptive, problem oriented
> action.[3]

Problem environments can be characterized as
structured or unstructured by the extent to which
programmed or nonprogrammed decision procedures
apply to them. Most interesting management decision problems appear to fall into the relatively

[2] Meador, C.L. and D.N. Ness, "Decision Support
Systems: An Application to Corporate Planning,"
Sloan Management Review, Winter 1974.

[3] Simon, H.A., "The New Science of Management Decision," New York: Harper & Row, 1960.

CHARACTERISTICS OF INFORMATION	PROJECT OPERATIONAL CONTROL	PROGRAM MANAGEMENT CONTROL	PROGRAM/PROJECT POLICY AND STRATEGIC PLANNING
Source	Largely Internal	→	External
Scope	Narrow	→	Very Wide
Level of Aggregation	Detailed	→	Condensed
Time Horizon	Historical	→	Future
Currency	Highly Current	→	Quite Old
Required Accuracy	High	→	Low
Frequency of Use	Very Frequent	→	Infrequent

Figure 5. Information requirements by decision category.

unstructured category. A typical attitude which
has prevailed in the past is that substantially
unstructured problems are either too trivial to
require decision support or so complex that it is
impossible to make decision support system technology relevant. A more useful design hypothesis is
that such problems are, in general, neither trivial
nor impossible; but they are difficult to solve.

It may be constructive to attack problems in
terms of subproblems and subdecisions that can be
supported rather than solved by a computer-aided
decision process. Aspects of the total problem are
sought where structure is recognizable, and the
computer is used to improve and mechanize those
aspects. Such subproblems may include comparison
operations, production of graphs or other data presentation, and arithmetic or primitive logical
operations. An understanding of common subproblems
helps separate real problems into more manageable
parts. The division of a problem into subproblems
accomplishes two distinct things:

1) Problems rarely look alike, but they
   often share common parts which can
   effectively be operated on.

2) The user is helped to structure his
   thoughts in an improved and helpful
   way by suggesting what kinds of subproblems might be particularly useful
   to attack.

These principles are important underpinnings to the
conceptual development and realization of decision
support systems.

Information Characteristics

Figure 5 delineates another important set of
design constraints for consideration in the development of a macro-engineering project planning and
control decision support system. It is important
to recognize that information characteristics vary
dramatically by type and scope according to which
decision environment is being addressed.[4]

---

[4]Gorry, G.A. and M.S. Scott Morton, "A Framework
for Management Information Systems," Sloan Management Review, Fall 1971.

	PROJECT OPERATIONAL CONTROL	PROGRAM MANAGEMENT CONTROL	PROGRAM/PROJECT POLICY AND STRATEGIC PLANNING
Structured	• Inventory Control • Cost Accounting • Quality Assurance	• Budget Analysis • Cost Forecasts • CPM/PERT Systems	• Economic Evaluation of Alternatives • Regulatory–Legal Impact Analysis
Semi-Structured	• Parts Production Control & Sched. • Work Force Assignment • Cash Management • Testing and Evaluation	• Technical Manpower Planning • Regulatory Agency Liaison • Research & Development	• Environmental Assessment • Political Response Analysis • Technology Forecasting • Social Cost/Benefit Analysis
Unstructured			

Figure 6. Macro-engineering management decision framework.

Earlier we stated that level of aggregation of information was quite detailed at the project operational control level but very condensed at the policy and strategic planning level. Required accuracy may be extremely high for operational control problems whereas a "reasonable guess" or a good estimate may be quite adequate at the strategic level.

From Transaction Processing to Decision Support

In the early development of the technologies of computer-based information systems, the field was led by an inward direction of concern with the hardware and software technology. This direction has resulted in a long series of orders of magnitude improvements in the processing efficiencies of new systems during the 1960's and 1970's. It can reasonably be expected that this progress should continue into the future.

There has also been a great deal of emphasis on the automation of relatively primitive, highly structured, previously manual tasks in areas such as personnel records, order entry, inventory control, and other accounting applications. Improvements in computer hardware and software technology have dramatically increased the efficiency of such transaction processing systems in recent years.

In Figure 6 we present a perspective on the macro-engineering management decision environment which we have characterized on the continua of structured to semistructured to unstructured decision problem type; and project operational control to program management control to program/project policy and strategic planning decision categories.[5] We have mapped into these two dimensions several specific tasks, applications, problems, and functions dealt with by macro-engineering technologists and managers. Inventory control, for

[5]This framework is a modified extension of early work on a general managerial information framework by Gorry and Scott Morton. See Gorry and Scott Morton, op cit., p.62.

The decision support system approach implies:

- Consequential problems in fuzzy, semistructured environments

- No programmed procedure applicable

- Requires managerial judgment, experience, intuition

- Tradeoff efficiency versus effectiveness

- Human interaction with computer

- Emphasis on supporting decision–making — not solving decision problems via the machine

Figure 7. Computer-based decision support system.

instance, can be viewed as a highly structured
problem to be dealt with at the operational control
level. At the other end of the spectrum we have
tasks such as environmental assessment, technology
forecasting, and social cost/benefit analysis which
are relatively ill-structured and are dealt with at
the policy and strategic planning level.

It is important to note in this framework that
most of the computer-based information systems
brought to bear on this problem context in the
past have been relevant principally in the upper
left quadrant of our figure. That is, the computer
aids have been of the earlier transaction processing orientation which was concerned with highly
structured operational control tasks such as inventory, cost accounting, quality control, reporting,
etc. The decision support system approach, which
we advocate here for macro-engineering project
planning and control, has the effect of moving the
context of concern from the upper left to the lower
right quadrant of the figure. There is an emphasis
on tackling some of the hard semistructured and
unstructured policy and strategic planning issues.
This approach requires us to start worrying about
and attending to the problem of improving the
effectiveness of organizational decision makers in
macroengineering contexts who are users of computer-based decision aids. This implies that less
time be spent worrying about improving the
efficiency of computer hardware and software. The
approach also leads us to emphasize a user-oriented
process for designing the interface between decision maker and computer. Many macro-engineering
technologists and managers may not be conversant
with a computer programming language. Thus, we
may, in many cases, be led to design a human-
computer interface with natural language characteristics. This will facilitate the iterative problem solving and exploratory data analysis processes
which seem to be important in rich, consequential,
semistructured problem domains. Figure 7 summarizes many of the characteristics relevant to decision support systems in this context.

In the next sections, we develop a project
modeling technique, the project management state
space approach PMSS, and describe a project planning and control decision support system called

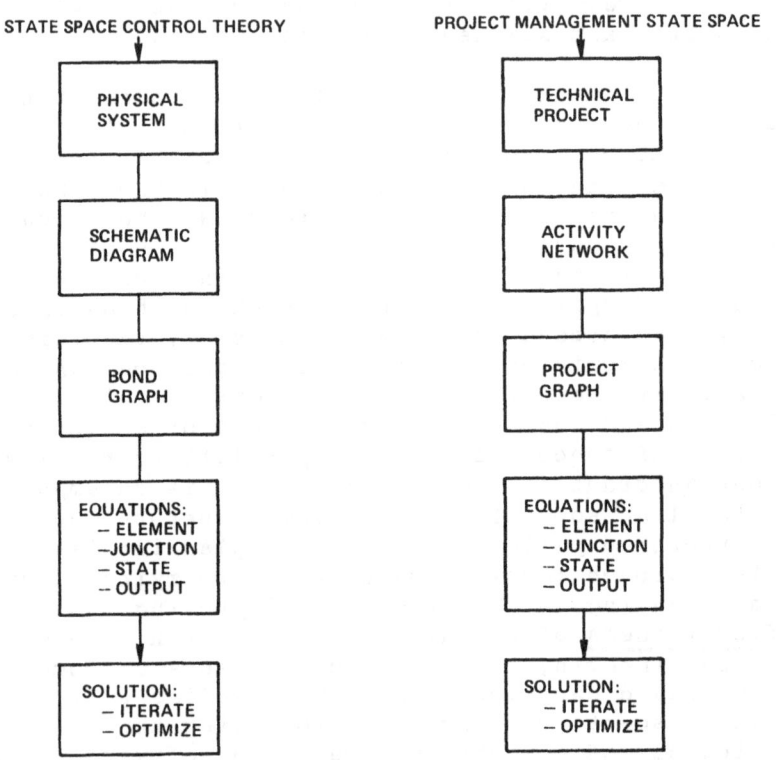

Figure 8.   Parallel development and implementation of PMSS.

MACRO-PLANNER which we are utilizing as an experimental test bed for the modeling technique.

The Need for a Model

The managerial decision-making process for macro-scale project personnel can be enhanced through appropriate utilization of decision support aids. One such aid being developed by the authors is a state space project modeling system. The motivation for developing this tool is the need for having available an accurate and relevant analytic representation of the project environment. This permits experimentation with optimization of project structure, analysis of sensitivities to and consequences of various perturbations, as well as other categories of "what if" interrogations. The computational and structural capabilities of state space control system analysis techniques have been found to be adequate for synthesizing macro-scale projects. The appropriate analogies for linking control system concepts with the project environment to justify adoption of current state space concepts are developed in an earlier discussion of this issue.[6] This is illustrated conceptually in Figure 8.

The graphical representation of important project variables has received considerable attention over the years. PERT(Project Evaluation and Review Technique) charting is still considered a valuable representation system. Here the project is represented as a series of line segments that correspond to initiation and completion of that activity. These activity line segments are interconnected in a sequential fashion that reflects their chronological ordering. This technique has been of most value in a project planning context.

A similar graphical technique was defined in reference 6. It was noted that the bond graph method for diagramming engineering systems had

[6]Parthé, Arthur C., Jr., Application of Modern System Analysis Theory to Large-Scale Engineering Development Projects, Second Lawrence Symposium on Systems and Decision Sciences, Berkeley, CA (Oct. 1978).

certain similarities to project (PERT) networks.
An activity graph project modeling technique has
been defined and is derived from both the PERT project activity network approach and the engineering
bond graph modeling technique. This facilitates a
convenient and tractable coupling of the project
environment with state space control systems
theory.

Project State Space Concepts

The concept of multidimensional space has
always had a particular intrigue and romance that
alluded to hidden promises. A project analyzed in
this framework is conceptualized as "originating"
from the origin of the requisite coordinate system.
As time elapses and resources are expended, the
project moves along a trajectory through "project
state space." The coordinates of this path are the
minimal set of parameters to uniquely specify the
project "location" at any given moment. These
parameters include project cost, schedule, technical performance and other related quantities of
interest. The difficulty in implementing these
concepts was the lack of a specific scheme for
tying this all together analytically. It is with
the marriage of state space control system analysis
and the conceptualized project state space that a
unified and tractable methodology has evolved.

Project Management State Space/PMSS

Project Management State Space (PMSS) concepts embody a methodology and analysis framework
that utilizes the resources of the digital computer. In a supportive role, PMSS can assist project management in planning and control decision
making. It deals with the semi-structured project management details that require management
judgement. PMSS is designed to enhance this process by making relevant information available on
an as-required basis. The analytical framework
for PMSS is derived from state space control system
theory.

System Modeling

System representation in state variable form
is given in the most general form by the following
system of equations:

$$\underline{x}(t) = \underline{A}\underline{x}(t) + \underline{B}\underline{u}(t) \tag{1}$$

$$\underline{y}(t) = \underline{C}\underline{x}(t) + \underline{D}\underline{u}(t) \tag{2}$$

where

$\underline{x}(t)$ is an n-dimensional state vector,
$\underline{u}(t)$ is an r-dimensional control vector,
$\underline{y}(t)$ is an m-dimensional output vector,
$\underline{A}$ is an nxn system matrix,
$\underline{B}$ is an nxr control matrix,
$\underline{C}$ is an mxn output matrix.

The added term $\underline{D}\underline{u}(t)$ allows for a direct coupling of the output and input. Equation (1) is a set of n first order differential equations and in control is referred to as the plant equation. Equation (2) represents a set of m linear algebraic equations and is referred to as the output expression. The $\underline{x}(t)$ vector has as components the variables that uniquely define the state space location and trajectory of the system. Analysis of the equations allows for observation and optimal solution of the control system being studied.

Optimal System Performance

In considering optimal performance it is necessary to specify a method for determining the quality of performance for the system under consideration. The current approach is to specify an integral performance index of the following general form:

$$PI \int_{t_i}^{t_f} L(\underline{x},\underline{u},t)\,dt \tag{3}$$

The intent here is to achieve optimal performance over the interval from t_i to t_f by minimizing or maximizing, as appropriate, the performance index PI.

System Modeling with Bond Graphs

The accuracy and ease of analytically assessing the performance and control of physical systems is dependent upon the specific modeling techniques employed. Methods that are either too simplistic or at the other extreme, overly complicated, are generally not productive for the analyst.

Table 1a.   Single port elements.

| Systems | Variables | | ONE PORTS | | |
	effort	flow	R resistor	C capacitor	I inertia
Systems	effort e	flow f	$e = Rf$ $\xrightarrow{\frac{e}{f}} R$	$e = \dfrac{q}{C}$ $\xrightarrow{\frac{e}{f=\dot q}} C$	$f = \dfrac{p}{I}$ $\xrightarrow{e=\dot p}_{f} I$
Mechanical	force F	velocity V	$\xrightarrow{\frac{F}{V}} R$	$\xrightarrow{\frac{F}{\dot X}} C$	$\xrightarrow{\frac{\dot p}{V}} I$
Hydraulic	pressure P	flow rate Q	$\xrightarrow{\frac{P}{Q}} R$	$\xrightarrow{\frac{P}{\dot V}} C$	$\xrightarrow{\frac{\dot P}{Q}} I$
Electrical	voltage e	current i	$\xrightarrow{\frac{e}{i}} R$	$\xrightarrow{\frac{e}{\dot q}} C$	$\xrightarrow{\frac{\dot \lambda}{i}} I$
Project (PMSS)	effort e	flow f	$\xrightarrow{\frac{e}{f}} R$	$\xrightarrow{\frac{e}{\dot q}} C$	$\xrightarrow{\frac{\dot p}{f}} I$
				$\xrightarrow{e=\dot p}_{f=\dot q} Z_i$	

Table 1b.   Multiport and source elements.

| System | TWO PORTS | | MULTIPORTS | | SOURCES | |
	Trans-former T	Gyrator G	Effort Junction 1	Flow Junction 0	Effort	Flow
Engineering	$e_1 = me_2$ $f_2 = mf_1$	$e_1 = rf_2$ $e_2 = rf_1$	$f_1 = f_n$ $\Sigma e_i = 0$	$e_1 = e_n$ $\Sigma f_i = 0$	S_e	S_f
Project (PMSS)	T $e_1 = me_2$ $f_2 = mf_1$	G $e_1 = rf_2$ $e_2 = rf_1$	A $f_1 = f_n$ $\Sigma e_i = 0$	O $e_1 = e_n$ $\Sigma f_i = 0$	S_e	S_f

In 1959 Henry Paynter of MIT devised a modeling system that was based upon the concept of energy and information flow. Named the Bond Graph method, basic energy dissipation and storage elements were defined and a graphical algebra developed. This technique incorporates the salient features of state space concepts. Properly augmented bond graphs permit direct writing of system state equations.[7] We have found them to be directly transferrable to the PMSS problem addressed in this paper. These thoughts will be developed further after a brief discussion of the family of system elements used in bond graph modeling.

System Modeling Elements

The bond graph system of modeling makes use of three fundamental single-port elements. These are the resistor (R), capacitor(C), and inertia(I). These elements either dissipate (R) or store (C,I) energy.[8] Refer to Table 1a for complete definitions.

The two-port elements transformer (T) and gyrator (G) are also used. Effort (1) and flow (0) junctions along with effort (S_e) and flow (S_f) source elements complete the family.[9] Refer to Table 1b for details.

The relationships between the various modeling elements and system variables are illustrated in Figure 9. The analogous PMSS variables have been added for clarity. These relationships will be discussed in more detail as to how they have been interpreted for PMSS analysis. A summary of the power and energy variables with their project analogies is given in Table 2.

Activity Graph Project Modeling

The simple circuit and bond graph of Figure 10 can be shown to exhibit several properties and characteristics that form direct analogies to the

---

[7]Paynter, Henry M., *Analysis and Design of Engineering Systems,* Cambridge, MA: MIT Press, 1961.

[8]Karnopp, Dean and Ronald Rosenberg, *System Dynamics: A Unified Approach*, NY, John Wiley & Sons 1975, pp. 13-65.

[9] Ibid.

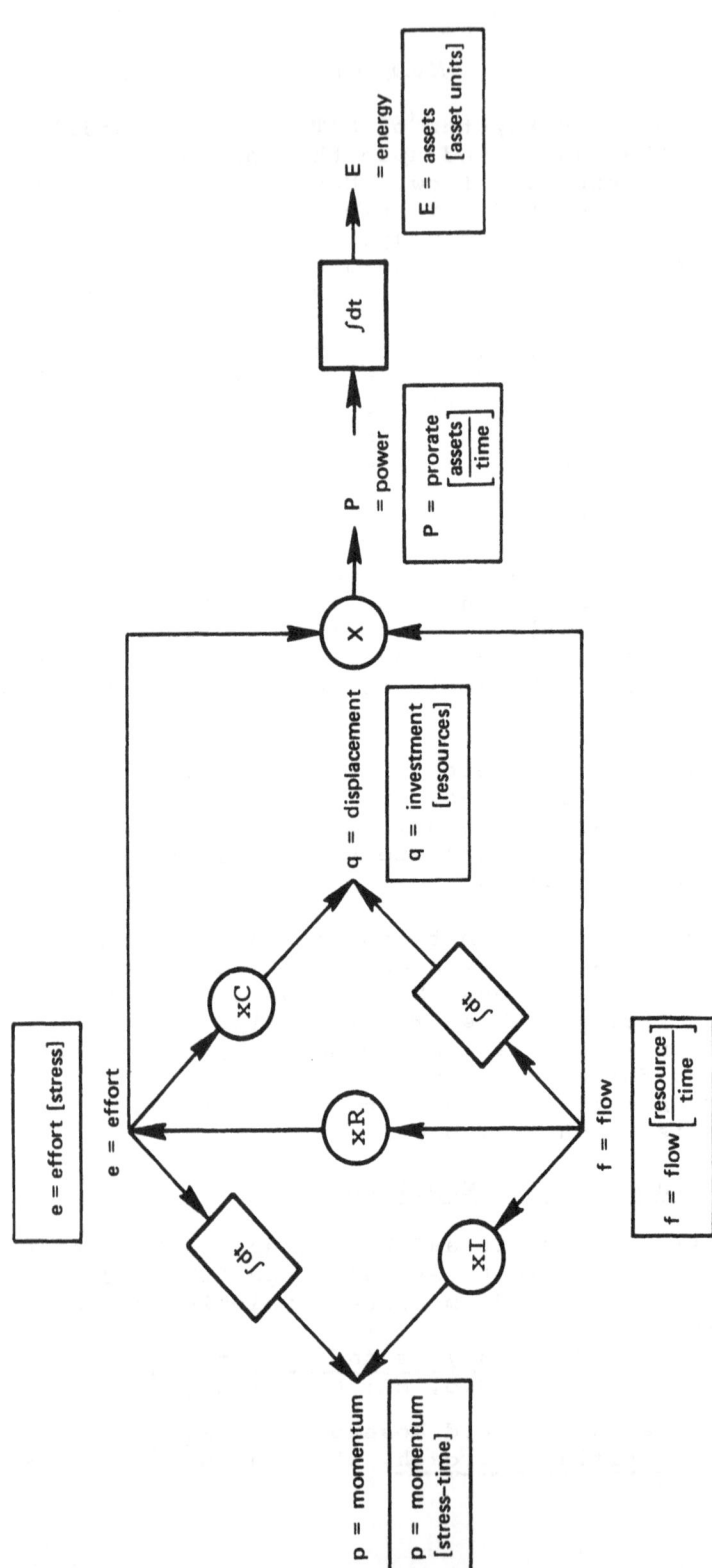

Figure 9. Relationships between modeling elements (R, C, I) and system variables.

Table 2. Power and energy variables.

System	Effort e	Flow f	Momentum $p = \int e\, dt$	Displacement $q = \int f\, dt$	Power $P = e \cdot f$	Energy $E = \int e \cdot f\, dt$
Mechanical	force F [newtons (N)]	velocity V [meters/s]	momentum mV [N-2]	distance X [meters (m)]	power F · V [(N-m)/s]	energy $\int$F·V dt [N-m]
Hydraulic	pressure P [N/m^2]	flow rate Q [m^3/s]	momentum P_p [(N-s)/m^2]	volume V [m^3]	power P·Q [(N-m)/s]	energy $\int$P·Q dt [N-m]
Electrical	voltage e [(N-m)/C]	current i [C/s]	flux linkage λ [V-s]	charge q [A-s]	power e · i [(N-m)/s]	energy $\int$e·i dt [N-m]
Project (PMSS)	effort e $\left[\text{stress} = \dfrac{t_n}{t_i}\right]$	flow f $\left[\dfrac{\text{resource}}{\text{time}}\right]$	momentum p [stress-time]	investment q [resources]	prorate P = e·f $\left[\dfrac{\text{assets}}{\text{time}}\right]$	assets E = $\int$e·f dt [asset units]

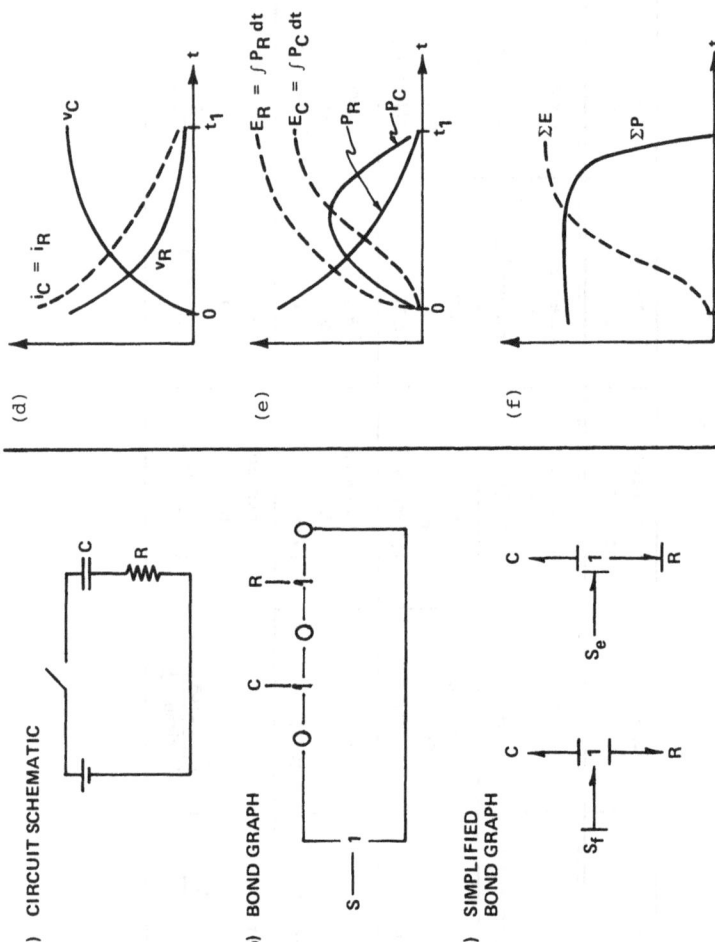

Figure 10. Electrical circuit analysis.

project environment. To illustrate this better, the partially simplified bond graph is repeated in Figure 11(a) (less ground return). Directly beneath it in 11(b) is a simple three-event activity network. These three events correspond with the three "0" flow junctions shown in 11(a). The interpretation is that the circuit element C and the power bonds joining junction numbers 1 and 2 represent activity 12. The dynamic characteristic of this "activity" is that of storing energy, or in PMSS parlance, creating "assets." In similar fashion, the circuit element R and the power bonds joining junction numbers 2 and 3 correspond to activity 23. The characteristic of this "activity" is that of dissipating energy. In the PMSS project situation this is interpreted as expending resources over a period of time without creating any retained assets.

Figure 11(c) illustrates how the "Activity Graph" is defined in relation to both the conventional project activity network and the bond graph. Preserving an activity network type format eases the transition to thinking in activity graph terms. The "A" junction (=activity) is functionally identical to the "1" effort junction in the bond graph. The dynamic characteristics of each activity are modeled by bonding the appropriate R,C and/or I elements to the A junctions.

One of the desirable features of PMSS activity graphs is the ability to analyze the flow of individual resources throughout a project. Thus, the dynamics of individual processes can be modeled more accurately. Assuming this is the case, then Figure 10(d), (e) and (f) are interpreted as showing the levels of effort e, flow f, prorate P, and assets E for the two activities illustrated in Figure 10(c). Accurate modeling of real project dynamics requires that the appropriate R,C, and I elements and their respective values be selected for each activity and resource flow.

An augmented PMSS activity graph is shown in Figure 12. A segment of a conventional activity network appears in 12(a) while an augmented PMSS activity graph is presented in 12(b). The difference in 12(b) is the addition of the activity effort junctions, A, and the dynamic element Z.

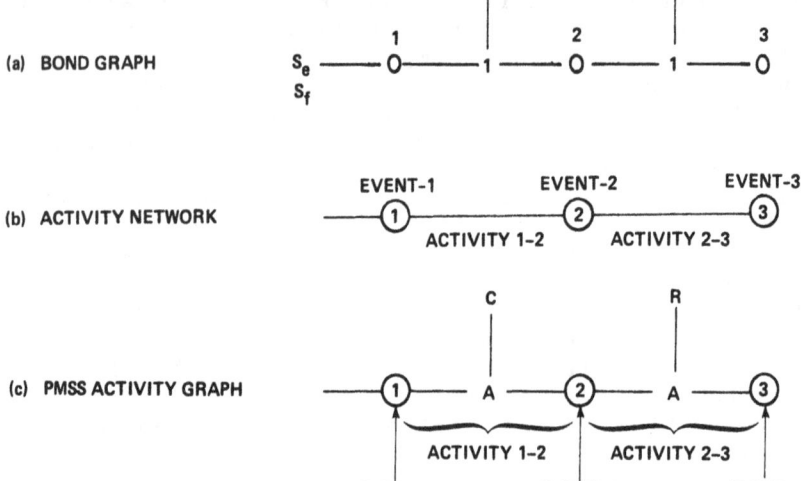

Figure 11.   Activity graph project modeling.

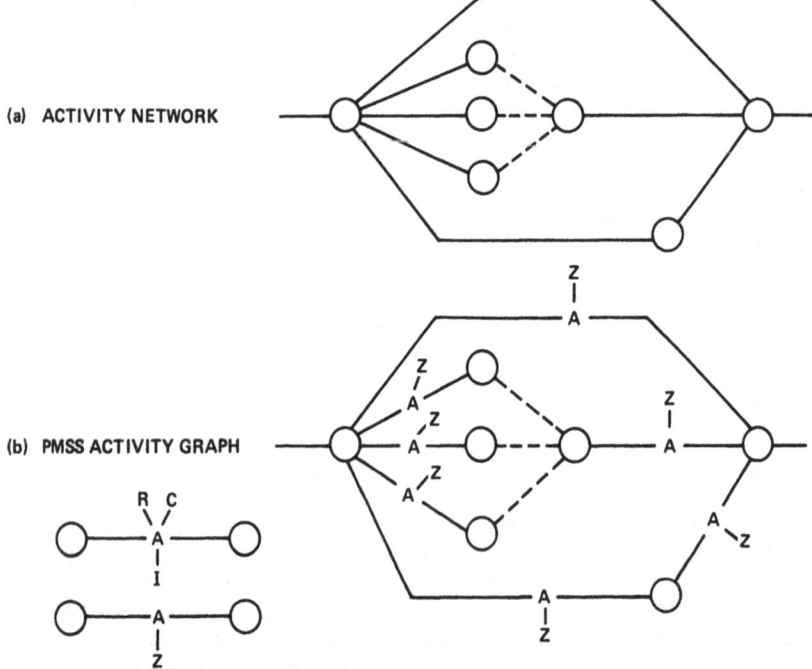

Figure 12.   Project activity diagram.

Rather than show R,C., and I separately, the three-component symbol Z (impedance) is used.

Not shown are the subelements of the "event" flow junctions"0". PMSS has three programmable subfunctions imbedded in each of these junctions. These subfunctions are:

1) Threshold detector that verifies when and where a preset level has been realized (e,f,p,q,P or E)

2) Event node that can be set to detect the coincidence of completion of other activities

3) Activity initializer that serves as both a source (S_e or S_f) and trigger for starting subsequent activities.

The addition of these subfunctions greatly enhances the analyst's ability to exercise the PMSS project model. It is this flexibility that allows for experimenting with project dynamics and evaluating various optimization strategies.

Operational Aspects of PMSS

The PMSS approach is currently being implemented in the MACRO-PLANNER project planning and control decision support system. In the following sections we develop an operational view of certain aspects of this support system. Figure 13 summarizes several key concepts and objectives of MACRO-PLANNER.

Modeling Structure

Project modeling with PMSS can involve three or more levels of structure. Going from the summary level down, these are successively called (1) macrostructure/space, (2) microstructure/space, and (3) nanostructure/space. From a conceptual viewpoint, this means that an activity vector in macrospace is represented by the sum of its components as defined in microspace. The same reasoning follows downward for microstructure vectors into nanospace. It is appropriate to think of these structural spaces as being successive hierarchical decompositions of each other.

MACRO-PLANNER characteristics:

- Emphasizes problems critical to large-scale developments
- Data base management systems and development tools
- Analytical tools
- Integration of available program management facilities
- Interactive interface for use by non-computer experts
- Computer graphics and report generation
- Exploratory data analysis
- Natural language and knowledge representation

Figure 13a.   MACRO-PLANNER: A macro-engineering project planning and control decision support system.

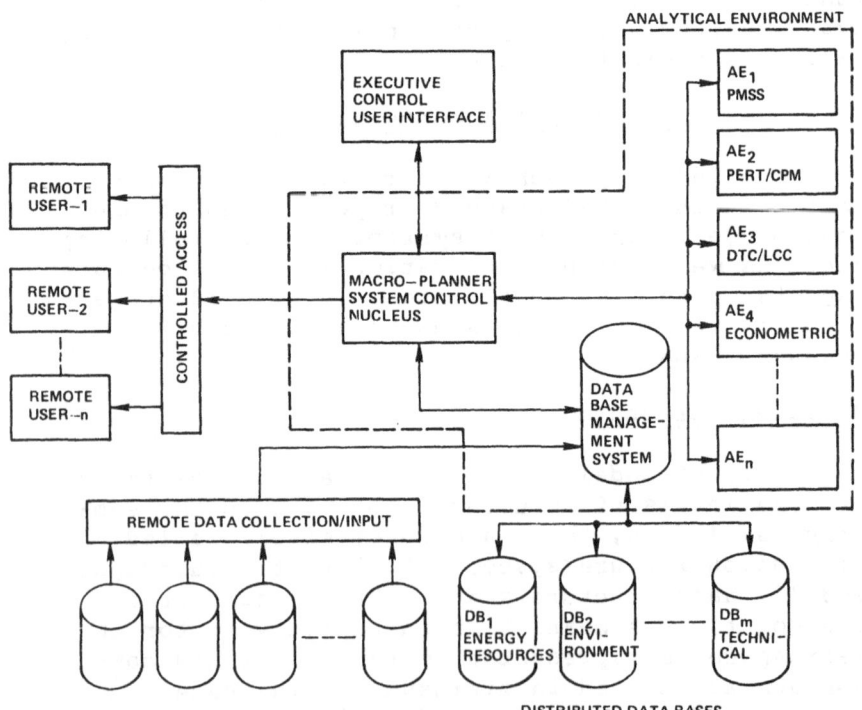

Figure 13b.   Macro-engineering program planning and control system architecture.

Data Display

Graphical illustration of state space, where the number of coordinates exceed three, has always been a challenge. Since the dimensions of Project Management State Space will exceed three for all real project models, an early effort was made to develop an effective graphical format, which is essential if one wishes to observe the status of schedules, cost incurred, resource utilization, technical performance, etc., at various points in a project.

The scheme that has been found to be most useful, to date, is illustrated in Figure 14. This represents a "slice" of the project state space at the area of interest. Two state points are shown, indicating the changes over time period Δt. The state points are spread out in two dimensions, each "vector" shown representing the magnitude of one state space coordinate (dimension). A significance to the angular orientation can be defined. Note that an observer in the plane of the figure, looking straight along the line joining the tops of the vectors, would "see" a state point.

The circles shown in Figure 14 represent the desired radial magnitude at time t and t+ Δt. All vector quantities are normalized with respect to them. The actual length of the activity state vectors are superimposed on this display. Vectors not long enough to reach the appropriate circle would indicate activity conditions of behind schedule, cost expenditure under budget, etc. Likewise, vectors extending past the circle indicate conditions of excess.

Additional interactive display formats are being researched for analyzing other aspects of PMSS project models. This interface with the system is critical since it could limit the analysts' effectiveness and efficiency to communicate in the modeling environment. Another graphic display is given in Figure 15, showing a typical "project state space trajectory."

Model Calibration

The PMSS modeling techniques described here require the analyst to synthesize an analytical

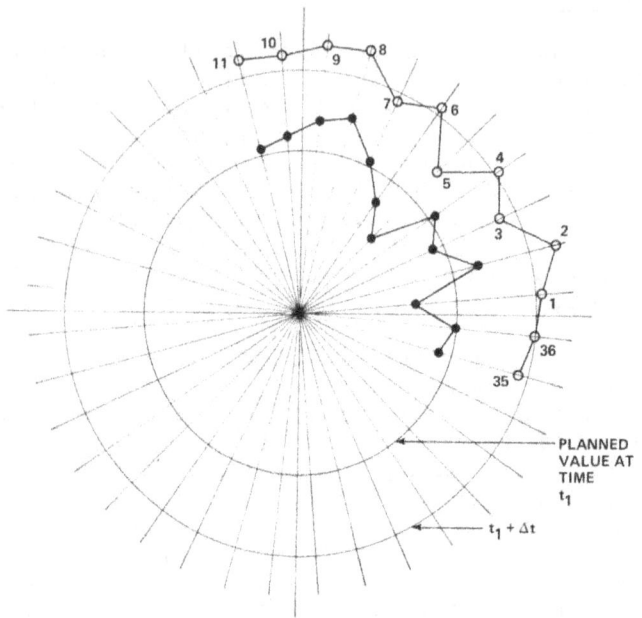

Figure 14.   Project trend display.

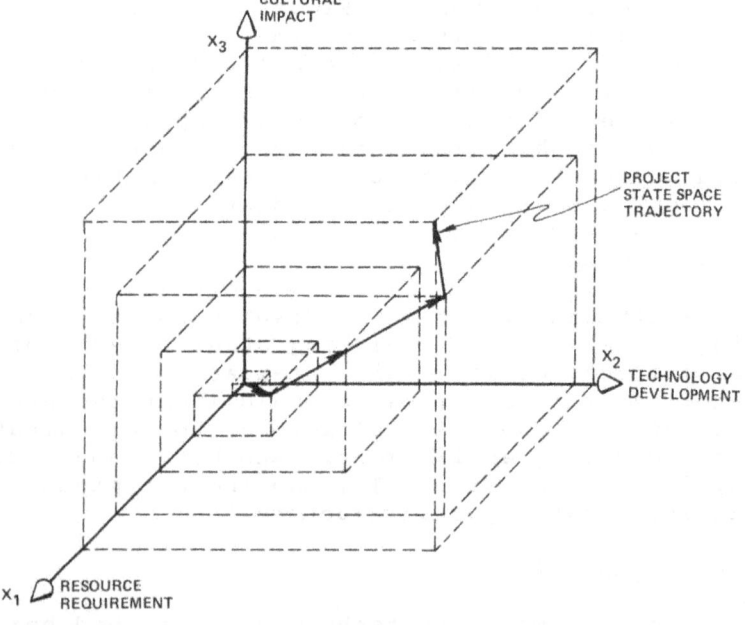

Figure 15.   MACRO-PLANNER/project management
state space.

representation of the macro-project environment.
As with any modeling technique, the result is an
abstraction of reality. The degree to which this
model accurately represents and anticipates future
status of the project depends on not only the
authenticity of the technique, but further on the
accuracy to which it is calibrated for a particular
project situation. With state space models, this
process is greatly facilitated when the state variables are observable and measurable.

The state variables that have been defined for
the PMSS model are "momentum" (p) and "investment"
(q). Momentum has units of "stress-time" and is
related to project schedule; investment has units
of "resources" and is related to the expenditure of
manpower, materials, etc.

The significance of having appropriately defined and observable state variables is that their
definition and quantification is sufficient to
adequately define the precise state space coordinates of the project trajectory through PMSS. In
state space control systems theory, a system is
said to be controllable if any initial state $\underline{x}(t_o)$
can be transferred to any final state $\underline{x}(t_f)$ in
finite time by some control $\underline{u}(t)$. For the PMSS
case, this latter property is interpreted to mean
that the state space trajectory (outcome) of a
project is predictable (controllable) using PMSS
techniques. This capability is of primary interest
and value with respect to the planning and control
functions of project management.

Knowing the values of the state variables for
the actual project over some period of time will
allow the analyst to derive what the appropriate
values should be for the model's activity impe-
dances. This is essential since the static and
dynamic performance characteristics of the model
are determined by the system element values and
their arrangement in the project model "circuit."

The requirement for adequate validation of
any model to determine the inherent accuracy and
range of application and thus predict reliability
is frequently not thoroughly addressed. This is
almost certain to lead to disappointment of the
user and discredit to the technique.

This eventuality is being guarded against in the development of the PMSS approach and the MACRO-PLANNER decision support system. Reliable and accurate historical data on several macro-scale projects is being used to verify that the PMSS model can in fact be validated. It is also building confidence in this new modeling technique and its application to the macro-scale project environment. It is further helping to mold the modeling structure as the technique evolves from the concept stage on out to a fully developed and deployable decision support system.

Conclusion

In this discussion, we have presented a framework which characterizes macro-engineering programs and projects from a managerial policy, planning and control perspective. From this framework, we have developed a multi-dimensional and relativistic project state space in which to portray analytic representations of macro-projects. We have derived in this state space a planning and control model, PMSS, which is designed to support the analysis and optimization of certain key project objectives and goals. The planning and control model is implemented and under study in a computer-based macro-engineering project planning and control decision support system called MACRO-PLANNER.

Values, Psychology and the Decision Process

Abandoned and undisturbed for centuries until its discovery by Yale University's Hiram Bingham in 1911, Machu Picchu justly deserves its title of "Lost City of the Incas." Perched on a saddle between the Machu (Old) and the Huayna (New) peaks, the urban site stretches over some 13 square kilometers. Its construction parallels that of most Inca "pucaras", or fortified city defense units. This important example of a macro-engineering project in the pre-Columbian Western Hemisphere includes a temple and a citadel that were once surrounded by terraced gardens linked by more than 3,000 steps. Eight-thousand feet above sea level and some 2,000 above the rapids of the Urubamba River, it is one of Peru's most impressive ruins.

Overview

Many questions can and should be raised about macro-engineering. What is our scale for measuring or comparing macro-projects? What is the motivation of the macro-engineer? What is the role of systems science (especially in relation to communications, transport and computer technologies) in modern macro-engineering? Finally, how does macro-engineering affect changes in human values as it inevitably will? Ramón Barquín explores these issues in some depth in order to suggest an appropriate conceptual framework for examining important social and cultural impacts.

5. On Large-Scale Undertakings and Human Values

Let me start by saying that I bring no answers, for I have understood my task to be the raising of questions.

Macro-engineering and human values...large-scale undertaking and human values. What do we mean by large-scale undertakings? Clearly, not every large-scale undertaking is a macro-engineering project, though every macro-engineering project is by definition a large-scale undertaking.

Engineering is the science by which the properties of matter and the sources of energy in nature are made useful to man in structures, machines and products. Macro-engineering...large-scale engineering....

Let's talk about scales....

Scales are for measuring...metric...relative dimensions... What must we measure, and what mark must we surpass before an engineering project can be prefixed with a macro? Is the scale in terms of total resources expended...duration of the project...manpower involved? Do we address the magnitude of the effort, or that of the results?

Large-scale, the dictionary tells us, also means: of wide scope. Are we concerned with the size of a structure...the complexity of a project.. the percentage of humanity benefited...or with still other characteristics of an endeavor? Maybe there is a direct relationship between many of these factors. Obviously there often is...

Our most frequent approach -- though not necessarily the best -- seems to be: let's look at cost...a money scale. "About a billion dollars, that's what a macro-engineering project is," a colleague recently volunteered. Are these Hong Kong dollars, U.S. dollars, or Australian dollars? Are they 1920, 1967 or 1978 dollars? What does this really measure? Are we to say that the Empire State Building does not qualify because it came in at a mere $42 million in 1931? In spite of its 60 miles of water pipes, 7 miles of elevator shafts, and the 7 million man-hours it took to complete? Or do we leave out the $650 million World Trade Towers finished in 1973, into whose 201 acres of rentable space, you can march every person in Denmark and stand them a yard apart?

Quite obviously, we can concert currencies and develop equivalency tables across time and space. Yet, even so, we should be very careful when we use a money scale. Can we accurately calculate the cost of building the Great Wall of China, which took place literally over two millenia, using enough material to build 120 copies of the Great Pyramid, and taking a toll of over 300,000 workers' lives during the process?

What's more, the Jubail complex in Saudi Arabia will cost about $20 billion before completion. This clearly qualifies it as a macro-engineering project. But what about efficiencies and inefficiencies? What about management and mismanagement? Would an unnecessary $10 billion cost overrun make Jubail a larger-scale undertaking than it already is?

Wide scope, complexity, orders of magnitude, challenge...in addition to cost. These seem to be the key ingredients. It was macro-engineering at the micro scale of the atom that launched us into the large-scale undertaking of space exploration.

Let us focus on the large-scale engineering of antiquity. Let us ponder the building of the Pyramids of Egypt...or the Grand Canal of China... or the great Mayan cities...or the majestic temples of Borobudur. Should not the scale be relative to the resources and the technology available to a specific society, in each corresponding historical period? Can we find a common scale to evaluate the engineering of Stonehenge and the building of Machu Picchu? Can we compare the development of the Concorde to the construction of the Inca bastion of Sacsayhuaman?

Would we want to?

Let's delve deeper into our subject...what motivates the engineer...the builder? Is it the same set of factors that motivate the macro-engineer...the large-scale builder?

What moved Imhotep -- macro-engineer, physician, and builder, later deified by the Egyptians--to design the Step Pyramid at Saqqara, model for the Great Pyramids of Giza? What made him take the step of constructing one mastaba upon another, breaking with all precedent and tradition? What motivates the macro-engineer?

Better still, really, who is the macro-engineer? Was it Imhotep, the builder; or was it Zoser his Pharaoh? Who conceived the grandiose scheme? Imhotep's name has reached us as the man who broke the molds at Saqqara and showed the way for the colossal works of Khufu, Khafre, and Menkure. However, there is no one name that passes down to us as having been the grand architect of these Great Pyramids. By this point, clearly, the projects had a loftier purpose: a house for all eternity...the proper abode for the immortal. For this, in the case of Khufu alone, it meant the skillful quarrying and masterful arrangement of over two and a quarter million stone blocks, some of them weighing as much as seven and a half tons.

The pharaoh was deemed to be immortal, and yet...might not the actual magnitude of the pyramid tomb that he envisioned and built be an attempt at finding personal reassurance of his immortality? A large-scale structure...a large-scale tomb...a definite attempt to reach out and grasp beyond our mortal world. Are not all large-scale engineering projects to a certain degree a reflection of this same urge?

Thus, we ask again: who is the macro-engineer? Is it the technician hard at work at the drafting table, or solving a series of differential equations? Maybe...but much, much more.

The macro-engineer appears to be the innovator, the grand director, the visionary, the master architect of real-life dreams. But most of all, the true macro-engineer is a man projecting himself to the beyond.

Man builds...man creates...it is in his nature. But when man decides to build in a scale which surpasses by far that which is common, he does so to reach out beyond himself. The justification for his actions may exist, the demand for his work might be there. Yet the grandiose solutions embodied in the macro-projects almost invariably entail this desire to transcend the limiting constraints of our nature.

In many cases,the undertaking might be dictated from above in one individual's search for immortality. In other cases, it might be the concerted action of a people who have decided to leave their mark upon this planet.

Ancient macro-engineering projects seem to have been mostly undertaken precisely by people who left their mark...empire builders in many cases. Was their willingness to engage in large-scale engineering projects related to their success in political, or economic development? Or was it the fact they were politically and economically successful that allowed them the luxury of these macro-endeavors?

Obviously, there are no clear-cut answers, but these two issues appear very related and they seem to reinforce each other. Nation building...empire building...they require commonality of purpose and organization. But, short of war, is there any other activity which achieves these two objectives better than a macro-engineering project? In fact, is there not an important parallel between these two large-scale undertakings in that an unusually large percentage of a people's resources are brought to bear on one sole activity?

It is reputed that at the peak of its construction, one out of every three males in China was working on the Great Wall. I wonder how many of the wars that China fought required a larger proportion of its men in the front lines.

Organization is an inescapable by-product. There is no way to engage in a rational scheme to manage the resources involved in building say, the Mayan city of Tikal, the Roman Aqueducts, or the pyramid complex at Chichen Itza without the development of an organization.

Organization implies some degree of specialization, and the skills developed in resource management and organization for these large-scale undertakings must have been extremely valuable in other national pursuits.

Furthermore, these huge landmarks of antiquity must have also provided a strong sense of national identification. Herodotus, a Greek, first tells us of the great colossi at Giza; but to the man of ancient Egypt, they must have been as much a symbol as the Statue of Liberty is to a New Yorker, or the Eiffel Tower to a Parisian, or to a Cuban the Morro Castle.

Let us come back to resource management and organization. The key element involved is information. Large-scale engineering projects imply the need to record, classify, store, modify and transmit large amounts of information. The demands for information handling capabilities, and man's attempts to fulfill these demands in the context of macro-projects were probably the principal factor leading to the development of an organizational infrastructure. Within these ordered schemes the transmission of information could take place in a more effective and efficient manner.

Communications and computation are critical engineering needs directly related to information. These are two important technologies almost always advanced by any macro-engineering project.

Technology is often defined as "comprising the practical media by which man interacts with his environment to meet his needs." More specifically, a technology is the set of disciplines, methods, techniques and supporting instruments which make up the process by which a tangible or intangible product is elaborated.

In any macro-engineering pursuit many individual technologies are utilized, according to any

definition. But in addition, the set of disciplines, methods and techniques which make up the
process by which a large-scale engineering project
is conceived, designed, planned, financed,imple-
mented and maintained in itself also features many
characteristics of a technology proper. Moreover,
because its components must interact in a system
fashion to produce its output -- the large-scale
engineering project -- it resembles a systems
technology.

Communications technology, transportation
technology, and computer technology are three examples of systems technologies. Today, any large-
scale undertaking must almost invariably utilize
these as instruments especially geared to handle
complex endeavors of wide scope, magnitude and
significance. This is the triad of the contemporary macro-engineer. They constitute the infrastructure needed for the macro-projects of our day.

But the computer and communications industries,
of course, have also seen their share of large-
scale engineering; and the results of many of these
projects have had an important impact on mankind,
and on human values.

Human values... Is this not also a part of our
topic? And we have not addressed it. Or have we?

What is a value? Kurt Baier tells us that it
is "an attitude for or against an event or phenomenon..." Kenneth Boulding defines it in economic
terms as nothing more than a "preference function."
In effect, a value can be just about anything that
we prefer in relation to something else. Thus a
brief list of common human values might include:
one's own pleasure, economic security, friendship,
privacy, nationalism, peace, rationality, power,
beauty...

We know that the values of a society play a
decisive role in human motivation. Values are at
the root of almost all social behavior. But should
not our main concern with respect to values probably be a dynamic one? Should we not be interested
principally with value change? With how values
change? When they change...and why?

Value change and technology seem to have a
very special relationship, though the mechanics of

their interaction are still relatively ill-defined.
For example, we have not yet been able to resolve
whether values came first -- as in Max Weber's
tenet that the Protestant work ethos gave rise to
modern technology and capitalism -- or whether it
is technology which, in the Marxist sense, determines society's values.

We do know that as a result of technology
certain values seem to be acquired or abandoned,
others, emphasized or de-emphasized. Alvin Toffler
has already said that value impact forecaster is
the profession of the future.

Do large-scale engineering projects cause
value changes? In many cases, probably yes. Did
not the completion of the Transcontinental Railroad
and the construction of the Panama Canal have an
important impact on some traditional American
values? Were not devotion to family life and
preference to live in one's home town somewhat
altered by enhancing the geographical mobility of
the average person? Has not the Manhattan Project,
which ushered in the nuclear energy age, had a
definite impact on man's peace values, as well as
on those related to prudence, and hopefully rever-
ence for life?

In our willingness to do battle with nature
and build a massive Grand Coulee or a Boulder Dam..
or construct the Itaipu by diverting the Parana
River for four years into a man-made cut...or put a
man on the moon...or harness the energy of the
atom...are we not in effect underwriting or
emphasizing self-reliance, self-fulfillment,
rationality, and power?

And, information...the principal commodity of
our time...has not its instantaneous transmission
through telephone and telegraph, satellites and
earth stations, radio and television, to all
effects shrunk our planet? And this has surely had
an important impact on many of our values.

And what about computers? Their capacity for
information-handling has pushed the limits of
knowledge much beyond the expectations of prior
generations. Random bits of data are usually
worthless; but organized information is knowledge,
and knowledge is power.

The marriage of computers and communications is allowing us already to develop a common infrastructure for the knowledge-based needs of society. "Think of information as energy," I'm sure you've heard before, and this is exactly right. The future should see each person's access and utilization of information sources greatly enhanced, and with this society's potential for optimizing resource management at the individual level.

We have considered large-scale undertakings, we have considered information, and we have considered human values. As I remarked at the beginning, I bring no answers, I hope I've raised some useful questions.

And, in ending, let me symbolically bring together the past, the present, and the future. For I have the uneasy feeling that through the ages the baton has been handed down from one macro-generation of macro-engineers to another in a continuous fashion: from the Pyramids to the Great Wall, Angkor Wat, Borobudur, and Machu Picchu; from the Taj Mahal to the Brooklyn Bridge to the Empire State Building to Hoover Dam; from the wheel to space travel, from the crank-shaft to computers and jet planes. And today, that baton is in the hands of the men of Itaipu and the TransAmazon Highway, of the planners of the "Chunnel" and the builders of the Jubail. Best that baton rest in steady and secure hands, for the macro-engineers of tomorrow, and mankind with them, will need it as they try to harness the tides and solar power, mine the ocean floor, capture icebergs, or energize the earth from a small black-hole.

Human values will change, there is no doubt. We trust that they will change for the better.

Notes

The quotes from Baier, Boulding and Toffler are from K. Baier and N. Rescher, eds., Values and the Future, The Free Press, New York, 1969, 527 pp.

Information on the Egyptian Pyramids is from A. Fahkry, The Pyramids, The University of Chicago Press, Chicago, IL, 1969, 272 pp.

Khufu, Khafre and Mankure are also known by their Greek names Cheops, Cephren and Mycerinus.

Other data on large-scale projects comes from the Encyclopedia Britannica, and the 1978 Guinness Book of World Records. Definitions are from Webster's New Standard Dictionary.

Overview

The immensity of macro-engineering projects
often arouses the emotions of both awe and fear in
man. Cynthia Oudejans Harris points out that part
of our fear arises because of the feeling that we
may not be able to say "no" to macro-projects which
technology makes possible but which are not funda-
mentally desirable. Macro-engineers and planners
must develop the ability to say "no" to some macro-
projects and become sophisticated about when and
why to accept or reject a project. To reduce our
fear of macro-projects, we must know that the project's purpose is sound, that the project is appropriate to its purpose, and that we can develop a
sense of "ownership" in the project through knowledge about it, participation in its planning, and
sharing in its use and development.

6. A Psychiatrist Reflects on Macro-Engineering

The title of this symposium, "How Big and Still Beautiful?" tells me that we are beginning to have strange feelings about "bigness." And I suspect that it is because of such feelings that I, as a psychiatrist, was asked to address this gathering. My colleagues and I are, after all, supposed to be specialists in strange feelings.

At the outset let me say that I believe that bigness is here to stay. A little later I shall comment on our gut and emotional reactions to the immensities amongst which we now live. But I wish to point out at the start that we have already intuitively begun to scale down our world: we no longer seek to exaggerate our mastery in our own eyes by creating immense public spaces, as we used to do; we will build no more Pennsylvania Stations, gorgeous though it was. And not only because it's too expensive: we don't need to any more. We have established ourselves as inhabitants of our planet. And in recent years we have found we need more in- timate spaces and smaller experiences to balance out our mastery over the sky and over the waters and over the earth....So the big law firms have begun to lower the ceilings in their old and dis- tinguished buildings; hanging plants are every- where; chamber music has become immensely popular; the brand new airport in Cleveland is filled with numerous spots where just two or three people can sit comfortably together; we have taken to vest- pocket parks; and we have taken to the mini-sport of jogging.

I mention these examples so we won't forget, as we talk about the big things, that we've recently taken to spending time, energy and money on small things, too. And that's very valid.

But back to bigness. We live in a world where the systems on which we depend are growing ever larger and ever more centralized and consolidated, whether or not we wish it. We could not even have come to Houston today were that not so. And, as you know, the 500 largest companies in the United States produce goods worth almost as much as those of the other <u>400,000</u> companies put together. Bigness just is the name of the game. We live on a planet which has turned MACRO in our lifetime, we live in a MACRO world.

What does this do to us at the gut level? How can we find our way amidst the immensities surrounding us?

It has been a privilege for me to devote some thought to the subject of immensity and the feelings it evokes in us. I shall then discuss some of the implications for macro-engineering which I see growing out of our response to the immense.

Let me start with something big <u>and</u> beautiful: Niagara Falls. Recently I visited them. I stood on the observation deck and the earth trembled slightly beneath my feet. My face was damp from the mists and I heard the great roar of the falling waters. I was filled with excitement. Then, as I neared the brim of the falls, where the water pours over in a two-foot sheet, I had a split second of fear. Would I be able to resist the urge to throw myself into it? Perhaps in a barrel, as so many before me? And would I, my tiny self, survive the challenge? Naturally, all of these feelings crowded into an instant, and I quickly, as our language says, "returned to my senses." I again became aware of the solid deck beneath my feet and of my own safety before this magnificent spectacle.

I mention these reactions of mine because I believe these emotions are fundamental to our relationship to the immense. This combination of the safety of the tiny spectator with the challenge of immensity, this mix of security and danger, of the sense of our own smallness and at the same time our wish to assert ourselves by confronting the challenge we experience -- all these feelings are parts of the instinctive exhilaration we feel when we confront immensity.

Are we enhanced or overwhelmed? We struggle between these two reactions. This ambivalence, I submit, constitutes the <u>emotional</u> challenge inherent in macro-engineering.

For what about the awe inspired by the immense creations of man?

Historically, human beings were first impressed by immense works of man -- not nature. The seven wonders of the ancient world, listed by Antipater of Sidon in the Second Century before our era, were not natural phenomena: they were all works of man. Like ourselves the ancients were impressed by man-made immensity.

When we feel dwarfed by modern man-made structures, our awe is tinged with the "pale cast of thought." "Who made it? Why did they? Who paid for it?," we wonder. And thus a touch of paranoia is added to our feelings of excitement and of fear.

In addition, as our capacity to create immense things has increased, we have witnessed so many of our solutions turning into our problems: we have seen industrial growth pollute our air and our waters, we have seen agricultural growth leach our soils, macrotankers pollute our seas. We have seen medical progress double the population of India within a generation -- now it stands at 640 million as against 340 million in 1948. And we have found our own powers to do the wrong thing no longer inhibited by the immensity of nature.

In consequence, we have developed a more intense uneasiness about our own power. "How big and still beautiful?" Can it be that Mies van der Rohe, the great modern architect, really <u>was</u> right when he said: "Less is More?" The scariness of "bigness", of the bigness of our projects, and, via them, of the size of our own power has begun to frighten us; we begin to think that maybe small <u>is</u> beautiful, if maybe less <u>is</u> more, and to wonder <u>if</u> big actually <u>is</u> inherently ugly. We wonder, too, if our own <u>power</u> is ugly. We experience a tinge of fear because we are, after all, so small. And a tinge of worry, because, after all, we <u>do</u> have

the power, -- and can we be sure that we will use
it wisely? And beyond that, we now know that it's
all up to us. Neither a god nor nature seems able
to control us.

But also we have new freedom.

A hundred and sixty years ago, in 1820, the
great German philosopher, Friedrich Hegel, wrote:
"The History of the world is none other than the
progress of the consciousness of freedom." Still
today there is much truth in his statement.

Freedom consists, first, in the availability
of choice, of outer choice; and, secondly, from the
emotional point of view, freedom means the capability of inner choise; emotionally, freedom is the
ability to say "yes" or to say "no" to the given,
outer environmental choice. For example, through
our institutions our society has given us freedom
of speech. That freedom -- which, of course, must
be constantly reconquered -- makes available to us
the freedom to speak. Then we must choose whether
to speak or to be silent.

So now, in Hegel's terms, we have made a piece
of progress. We are conscious of possessing a new
freedom: we have become outwardly free to build
great projects, great tunnels and bridges and air
networks, to build new cities in the jungles and
deserts, to build great islands in the seas and to
begin to create settlements in space.

What can my discipline offer in all this?

In a free society, such as our own, psychiatry
is intimately connected with freedom. It is my
opinion that all of our best psychiatric activity
is devoted to helping people develop the capability of choosing and of acting on, living out of,
informed choices. "Mental health," as it is
euphemistically called, is nothing more than the
ability to make free and responsible choices and
to live out of them.

(Of course, freedom is also onerous and
difficult -- which is why my profession exists.
Sometimes we want very much to have some "Big
Brother" relieve us of our freedom, to remove us
from the responsibility for our lives.)

So how can we as engineers, planners, informed citizens best live with our new macro-freedom? How can we be more enhanced and less overwhelmed by it?

The first thing we need to do is to develop our own **sense of choice**. We need to **know** in our very bones that, no matter how seductive a given project may be, we are at least **able** to say "no" to it as well as to say "yes" to it.

You all remember Hillary's famous response when asked why he bothered to climb Everest: "Because it is there," he said. We need to embrace the freedom to say: "Well, no, I won't climb it, --even though it **is** there;" and to say: "Even though this project may be the biggest challenge I will ever face, we will not build it. No, this project should not be built." For remember -- in case you happen to be into challenges -- the biggest challenge may actually be to say "no" to the "biggest challenge."

For, if we **must** say "yes" because this is our great challenge -- or simply because it is so marvellously exciting -- we are not truly free. To be truly free, we must have "no" as well as "yes", "yes" as well as "no" in our repertoire.

Of course, both our "no's" and our "yes's" -- and also our "maybe's" -- must come out of our own very best knowledge. This is not so simple as we perhaps once imagined. For it may very well be that, even with the aid of our most trenchant data and our most advanced computer techniques, we will still be unable to foresee all that may come about because of an immense project; furthermore, the data and the circumstances have become so complex that it takes more than one mind to encompass the realities being dealt with. Each individual can only comprehend **parts** of the whole. That is scary, too. Who **can** really know? How **can** we, as individuals or as groups, act in a truly realistic and responsible fashion?

Possessing the emotional capability of saying "no" or of saying "yes", is a beginning. Becoming truly sophisticated about **when** and **why** to say "no", **when** and **why** to say "yes" are skills that we will need constantly to refine.

So far I have been speaking chiefly about us in our capacity as decision makers or in our capacity of informed citizens.

Now I want to turn to the impact of man-made "bigness" on us as consumers, as spectators/builders/users of the great macro-engineering projects. In and of itself, immensity awakens our awareness of our tinyness -- as I have already mentioned, we easily feel overwhelmed and powerless and a little paranoid when we come upon some immense structure that we didn't even know existed. And if we then imagine that "they" built it with some vaguely thoughtless or perhaps deliberately evil purpose in mind, without consulting us about our needs, wishes, desires -- for sure, this immensity is "bad".

I had some such feeling when recently -- quite by accident, just following a country road along the shore of Lake Erie -- I stumbled upon the nuclear power plant, now being built in Perry, Ohio. There were no signs, no notices, nothing -- only cultivated fields on either side of the dirt road. And suddenly there it was, immense, with a big fence and guards and lots of activity. I hadn't really been quite aware that it existed in my own back yard. I was left with a vague feeling of having encountered something evil. In my head, at least, this place belonged to "them" and not to "us". My guess is that each of you has had at least one similar experience within the past few years.

So again back to Dr. Davidson's guiding question: "How big and still beautiful?" I don't think this question can be answered in terms of meters and kilometers. Beauty, after all, is in the eye of the beholder. What sorts of things do we as consumers, users, spectators of macro-projects need, in order to be able to welcome a macro-project as "Wow! Beautiful!"

There are at least three things we need to be sure about if the big things are not to make us paranoid but rather to be beautiful to us.

The first thing is that we need to be sure that the purpose for which the particular project is developed is a sound one. To lend a big project

our support and not simply to resent it for its
very size we need to feel that it is a "good thing"
and that those who planned and developed it had
their heads about them as they planned: that our
needs for a clean and healthy natural environment
were attended to, that long-term effects were considered, and so forth. As far as I am able to tell,
precisely this sort of attention to purpose went
into the building of the immense new polders --
540 square miles of man-made land, so far -- on the
bottom of what used to be the Zuider Zee in Holland.

Once we know that the purpose is a good one,
the second thing we need to know is that the immensity of the project is appropriate to its purpose
and to the available resources. The Dutch wouldn't
empty a bay to make a single acre of land; only
hundreds make any sense. And on the other hand,
Proposition 13 in California has turned several
immense projects into white elephants, as for
example the gigantic, useless highway interchange
-- disconnected from any other highway -- to the
north of San José. The immensity of that project
became -- and was that foreseeable? -- inappropriate to the available resources.

And the third thing we need to know -- if an
immense project is to be big and still beautiful in
our eyes -- is that it belongs to "us" -- not to
"them". I had no such feeling, peering through
the fence at the nuclear plant. It surely didn't
belong to "us".

How do we get that feeling of ownership? And
what can we do to foster it in others? I want to
discuss this process in some detail. Daniel
Yankelovich, the economist, has pointed out that
one of the human factors contributing to the
present surge of inflation is what he calls the
"philosophy of entitlement"[1]: we have begun to
feel entitled to care by our government when we are
unemployed, sick, old or otherwise plagued. I
wish to suggest that a "philosophy of ownership" is
one of the antidotes to the "philosophy of entitlement" and that we need to pay a great deal of

---

[1]Cited by Robert L. Heilbroner, *The New Yorker*,
August 28, 1978, p. 57.

attention, <u>particularly</u> in the area of macro-engineering.

Not a great deal is known about this important process, but let me share the little that I do know.

Traditionally, we think of "ownership" in financial terms. But to own something financially may or may not help to confer the sense of empowerment, of ownership, to which I refer here. Let me give an example. We talk about "<u>our</u> government" and, although we are weary of paying taxes, we don't feel as if we <u>own</u> the government. But it is "our" government, nonetheless. This is the sense of ownership to which I refer.

As to the process of gaining ownership -- I see three major activities through which we may grow to have a sense of an immense project belonging in some way to us. Doubtless there are other activities as well. The three that I see are: firstly, by becoming knowledgeable about the project; secondly, by having a voice in the decision-making process involved in its planning and consummation; and thirdly, by <u>using</u> and caring for the project.

As to the first way: becoming knowledgeable. Becoming knowledgeable is a primary way in which we gain a sense of ownership about virtually everything and anything. Our knowledge about our scientific fields gives us our capacity to participate in them and hence our sense of ownership of them. And if I know enough about my refrigerator to be able to fix it, I have a much stronger sense of ownership than if I have to call the repairman. Transferring this principle to macro-projects, it is clear that public education about the thinking that went into the planning, about the opportunities and the dangers and the drawbacks involved, is of the essence. And, for this nation of do-it-yourselfers, models and technological data of all sorts should be readily available to everyone. Further, in my own opinion, significant funds should be set aside for just this purpose of informing and educating the public. For, let me stress again, it is not only that data is being passed about: in addition our own sense of ownership, of empowerment, is being nurtured.

The second process by which we may gain a
sense of ownership is by having some say in the
planning and development of the big projects.
Everyone whose life will be intimately affected by
any big project should be involved as much as
possible in the process by which the plans become
reality. At the very least, if your life is going
to be affected, you must have an opportunity to
have your say about the big structures that are
going to have an impact on you. Just having your
say will help you feel that in some way the project is yours, that you, too, have some power over
the outcome -- even if your wish not to have the
new highway involve the destruction of your house
cannot be granted. All this, of course, provided
that you are heard out in good faith. If you are,
having the opportunity to have your say accomplishes a second thing, too: you begin to
believe that someone at least gives a damn about
what happens to you and to your neighbors. Really
to give us consumers a real chance to influence the
big plans, or minimally to have our say about what
is being done to us and to our environment, is an
extremely time-consuming and difficult work; but it
is an essential part of emotionally detoxifying
immense projects. It is related to the kind of
investment of time and energy that a surgeon must
make in talking with his patient about the operation he is going to perform. An outstanding
example of this kind of investment of effort (and
money) was the work done recently by Dalton, Dalton,
Newport under the leadership of Engineer Mel Lehr,
in soliciting public input into the Baltimore
Regional Transportation System Project. Lehr
stresses that this was not P.R. work but rather
that genuine public involvement in the planning
process was a major corporate objective.[2]

A third way in which we come to have a sense
of empowerment and ownership is through using a
project and, perhaps, even helping to take care of
it. We have all had experiences of an enhanced
sense of ownership. Our old blue jeans: they have
been through so much with us, they really belong to
us! And our house is really "ours" -- even if the
bank owns most of it -- because we have invested
ourselves in fixing it up, turning it into "our"
place, putting our own stamp on it. You can all

---

[2] Personal communication.

think of other, perhaps better, examples of how you
have come to have an enhanced sense of ownership
over something -- perhaps a park, a restaurant, a
particular street corner.

On a larger scale, we can say that the biggest
macro-engineering project of the past two centuries
has been the settlement and development of this
country. We have used this country, and in some
strange way it has come to belong to us. We and
our forefathers explored it, developed it, fought
over it, built our homes and our factories and our
public buildings on it, and we are here in Houston
today trying to help make it a better place -- we
have used it and it has become ours. We have a
sense of ownership. Even the trash cans admonish
us: "Help Keep Our City Clean."

I have been talking, you recall, about three
possible activities by which we may come to have a
sense of ownership: through getting knowledge,
through having a voice, through being a user. And
I said that our sense of ownership of macro-
engineering projects is important for our sense of
ease with them, is important if we are to find
them beautiful.

I want to stress that this really is a question of pronouns: if "they" own it and run it, the
immense project stands a good chance of being
viewed as potentially dangerous at best, as evil
at worst; but if "we" have a sense of ownership of
the great project, if it "belongs to us," the
chances are that it is a "good thing," that it is
"beautiful."

Therefore, as consumers, we must ask ourselves: "Does this belong to us? Do we own this
at all?" And as planners and engineers and executives we must ever attend to the process of giving
ownership of great projects to those who use them
or on whose lives they otherwise have impact.

We need to learn a great deal more about the
process by which we begin to say "we" and to take
ownership. Many issues of trust and honesty, of
information and judgment, are obviously involved
here. The whole thing is not a simple matter. I
have tried to suggest some ways in which our sense
of ownership is evoked. I am rather sure that we

cannot be manipulated into a fake sense of ownership. It is my own fervent belief that our language is so powerful that we cannot be maneuvered
into saying "it's ours" if, deep down within
ourselves, we feel it belongs, in actuality, to
someone else.

There is a further aspect to all this which I
must mention in closing which is of particular
importance to macro-engineering projects which are
of inter-cultural, international scope. The things
I have been saying apply, I am convinced, in our
own American culture and substantially in most
Western cultures. However, I would not venture to
say I know much at all about how an African or an
Asian comes to have the kind of sense of ownership
which I have been describing -- or even whether
such a sense of ownership is appropriate, desirable or possible. Perhaps a quite different group
of responses to immensity govern, in another culture. For instance, Akira Kawada, the Japanese
psychiatrist, recently wrote about "the unique
usage of personal pronouns in Japanese." In the
Japanese language the first person is expressed by
various pronouns depending on the relationships
with persons addressed. For example, the formal
word for "I" is "watakushi," but men often use
"boku" among friends and "ore" with their wives,
intimate friends or subordinates. In other words,
the "self" of a Japanese is defined by his relations with others.[3] Even from this limited information about the Japanese language it is clear that
"ownership" in Japan must have a very different
feel to it than our own. How their use of "we" and
"us" differs from ours, I have not had opportunity
to discover. But I am sure that Japanese issues
around "How big and still beautiful?" are different from our own. And that in still another cultural sphere they would be different again.

When working abroad, it certainly behooves us
to sharpen our own awareness to the ways in which
foreign perceptions, even of such apparently simple
matters as time and space, differ from our own.

In my remarks I have tried to explore our
possible inner feelings when we experience

---

[3]Akira Kawada, VOICES: Journal of the American
Academy of Psychotherapists, Vol. 14, No. 2, p.42.

Figure 1.   Early photo of Henry Wingate, the
author's great-grandfather, who was an early
settler in the state of Iowa.

ourselves as macro-types; I explored some of the
ways we feel when we experience ourselves as micro-
types confronting the works of macro-types, and I
pointed out some ways in which macro-things may
appear less sinister and more beautiful to us.

In closing, I wish to review my main points.

1)   Less COULD be more.
     Mies van der Rohe was the first to state that
     "less IS more." As macro-types we need to
     recall that less MAY be more and so to culti-
     vate the freedom to say "no" as well as "yes"
     to immense macro-engineering projects.

2)   Immensity fills us with awe.
     Awe is a mixture of the excitement of challenge
     with the fear of danger.

3)   Awe easily becomes paranoia.
     As a micro-type, it was easy for my awe to
     become paranoia when I stumbled upon the new
     nuclear power plant now being built near
     Cleveland, Ohio.

4)   It becomes ours when we:
          -know a lot about it
          -have made our own decision about it
          -use it
          -help care for it.

     Henry Wingate, my great-grandfather, in his
     hand the pitchfork with which he helped to make
     arable and to settle the state of Iowa. You
     can see that he knew a lot about it, had made
     his own decision about it, had used it and
     helped to care for it. And you can see that he
     felt good about all that.

5)   If it's "ours" it's BEAUTIFUL!

6)   Different cultures perceive things differently.
     In the beginning, only the Europeans wanted to
     climb the great peaks of Nepal. These moun-
     tains provided no challenge to the people who
     lived near them -- they perceived them
     differently.

Overview

The principles of social psychiatry, whose application has increased the capacity of both individuals and groups to deal with stress, can help to reduce the adverse social effects of some macro-engineering projects. As explained by Hans R. Huessy, these principles assert that any well-functioning social system provides the following attributes: 1) leadership, 2) social support, 3) opportunities for constructive participation, 4) positive expectations, 5) adequate information, and 6) anticipatory guidance. In many cases, technological advances have removed these attributes from our society. Social engineering on a macro-level is needed to correct the large-scale imbalances which exist. And successful macro-engineering of any kind requires the development of a "macro-morality" in which responsibility for the results of macro-engineering is not shunted aside.

7. Macro-Engineering and Social Psychiatry

Our topic, "How Big and Still Beautiful," was meant to question the effect macro-engineering projects have had on their beneficiaries, human beings. Unfortunately, the answers provided in our discussions were not wholly adequate.

Perhaps we should look at some early macro-engineering projects and see what effect they have had on people. One of the first big apartment houses built in Marseilles had a playground on its roof. The building made the architect very famous, even though the playground invited disaster, since every pervert who wanted access to children could commit crimes unobserved by any other adult. A garden apartment on the roof solved the problem. We have always had difficulties in our large apartment houses with the hallways, which are havens for crime. I have never understood why hallways are built without any windows from the adjoining apartments. Anyone in a hallway would have to worry about being observed—an unattractive nuisance, certainly, in the eyes of the criminal. Habitat at the Montreal World's Fair came to grips with this; hallways were only on every fourth floor, exposed. And the space they occupied could be used well by the occupants.

We don't have to look too far back to see even large-scale mistakes made in our attempts to provide efficiency and good living through macro-engineering. This country built huge subsidized, low-cost housing projects for multi-problem families. We have had to abandon many of these projects because, to our surprise, hundreds of multi-problem families housed in one area didn't

stop having problems. From all we know of how
groups work, we should have been able to predict
this, and should instead have encouraged subsidized apartment programs for multi-problem families
placed among non-problem families.

The construction of large planned communities
like Reston, Virginia, and Columbia, Maryland, outside Washington, D.C., and some new towns built in
England, also should teach us some lessons. Much
effort was spent in Reston and Columbia to provide
adequate social organization, but it did not work
out. One reason was that the population attracted
to these cities at the beginning was transient,
with an average stay of one to two years. Second,
most of the mothers worked, so the children came
home to uninhabited neighborhoods. And third, no
older generation was present. The English towns
provided their own industries for employment,
giving greater population stability, but also
suffered from having an absent older generation.
Consequently, these big projects, very carefully
engineered to accommodate workers and their
families, lacked social structure. One curious
aspect found among these inhabitants was an over-
use of medical facilities. Apparently the doctor's
waiting room, not the neighborhood or work room,
was a good place to meet people, even though an
expensive one.

How can we house or provide work space for a
great number of people and still sustain a decent
way of life for them? Dr. Cynthia Harris emphasized our need to develop a sense of ownership in
the projects' beneficiaries, and provide the public with adequate information about the projects.
She suggested too, that we should allow them to
feel that they had participated in the decision-
making about the project.

This leads me to think that perhaps all
designers of macro-engineering projects should
keep in mind the principles of social psychiatry,
whose application has increased the capacity of
both individuals and groups to deal with stress.
These principles state that any well-functioning
social system provides the following attributes:
1) leadership, 2) social support, 3) opportunities
for constructive participation, 4) positive

expectations, 5) adequate information and 6) anticipatory guidance.

Let me give brief examples of these attributes.
We know that good leadership in a military unit or
an institution makes for healthy functioning. When
leadership is poor, even basic functioning deter-
iorates. When a military unit has an unusually
high psychiatric casualty rate, poor leadership is
the most common cause. The second attribute is
social support. We send a Red Cross wagon out to
disasters not because the people need coffee and
doughnuts, but because the presence of the Red
Cross is an indicator that the larger world cares
and is supportive.

A very obvious example of the third attribute-
opportunities for constructive participation --
comes from an experience at the end of World War II.
The Japanese were using suicide planes and when one
of these planes hit a ship, its bomb didn't always
go off. The plane's gasoline, however, might be
spilled around an anti-aircraft gun on the ship, so
that the gun crew could not fire the weapon. Due
to some Navy regulation, the crew had to stay at
the gun, even though it could not be used; and the
psychiatric casualty rate would be as high as 60%
to 70% because planes were coming at them and they
were helpless. There is burgeoning literature
showing that helplessness in many forms has deleterious effects on human functioning. (The regulation was soon changed.)

The importance of the fourth attribute, positive expectations, is well-known in schools, of
course, and I commend to you the literature on the
impact of "labelling." The fifth attribute, providing adequate information, is self-explanatory.
Finally, with the sixth attribute, anticipatory
guidance, we prepare people for stressful situations which we know they will encounter. A trial
run increases their ability to cope.

Buildings, bridges, tunnels and roads are not
the only kind of big projects. Creating social
organizations can also be a big project. We might
use William James' proposal for a moral equivalent
for war, or my father's ideas on planetary service
as an example. In essence, all young people would
give two years of service to their country, to do

the things which need doing but will never be
economically feasible or offer careers. This could
mean our providing genuine caring to the chronic
physically- or mentally-handicapped, or a plentiful
supply of teacher-companions to children from
deprived backgrounds. They could clean up the environment or engage in military service, the Peace
Corps or any other similar activity. All these
tasks need to be done, but do not offer lifetime
careers. Intensive caring of the disabled
requires people. But it is so exhausting that most
of us cannot do it except for a time-limited period,
six months to two years. There is no career in
being a Big Brother to a needy adolescent; there is
no career in enriching the lives of the chronically
handicapped, but it needs doing. The Peace Corps
and VISTA are small-scale attempts to offer opportunities for service, and the Civilian Conservation
Corps of the 30's served some of these same purposes.

We tend today to assess the needs of a group,
and then try to meet that need by paying some professionals to do so. Such programs tend to have
little stability through time. We would recommend
more "complementary" programming in which two
groups meet each others' needs. For instance, the
Foster Grandparents' program meets the needs of
both the children and the grandparents. Likewise,
while the Peace Corps met a need in underdeveloped
countries, it also met the needs of young Americans. The role of the professional is to supply
the framework only. I believe complementary programming could be developed on a very big scale.

Universal Service would bring together young
people from different classes, and backgrounds,
and help them feel that they had a personal stake
in their country. The youth who feels useless and
unwanted would experience being needed. The appeal
revolutionary activity has occasionally had for
these young people would be diminished and capitalism, the goose that laid the golden egg in this
country, would gain more time to solve its human
problem.

Along the same lines, if we can build space
platforms to generate electricity, we should be
able to link a big social project with a big
physical project. We need to build adequate

housing for low-income Americans. There may not be
much money to be made in such an undertaking, but
it may be crucial to our survival. Could we use
some of our unemployed workers for this?

We need macro-engineering in social engineering, but we must not let ourselves be dazzled by
either the size or innovativeness of a project.
In programs for humans it is often best to start
small. The recent change-over to modern math in
public education is a good example. Social engineers in ivory towers decided that by teaching
arithmetic as though it were algebra, children
would be better prepared to deal with abstract
mathematical concepts they would encounter later.
Up to that time, one of the few school subjects in
which poor children did well was arithmetic. By
turning the arithmetic into early algebra, we
guaranteed that the poor no longer could have even
this achievement. Instead of trying the new math
in a few school systems to see what would happen,
it was pushed onto the whole country because it
seemed so reasonable. "Reasonable" or Cartesian
thinking is excellent in the science of things, but
it does not work when applied to human situations.
Thinking should not precede experience, but must
follow it; and however reasonable something may
look, we will not know how it affects human beings
until we try it. Rational thinking cannot deal
with values or morals, which are at the heart of
so many human dilemmas.

Another recent example of a disaster in social
macro-engineering is the American love-affair with
psycho-analysis, a theory based on non-verifiable
hypotheses, which implied both the prevention and
cure of psychiatric disorder and problems of living. A society priding itself for its rational
thinking embraced a completely irrational enterprise. The theory speculates about causes of
psychological disorder. Treatment is then based
on the same theory and aimed at the supposed cause.
Medicine does not treat causes. Knowing the cause
of an illness makes possible its prevention but not
necessarily its treatment. Knowing the treatment
of an illness does not help one to prevent it. A
treatment should be judged by its efficacy and not
by its theoretical fit. If we were to treat a
smoker for cancer of the lung, treating his smoking
would be an inadequate intervention. If I were

drunk, fell into an ice-cold river and developed pneumonia, the mental health professionals would propose heating the river. I would want a shot of penicillin. With this kind of intellectualization so prevalent in mental health, it is no wonder that no one ever realized that Freud's description of how children develop was unrelated to how children were actually raised in his day. In Europe, all homes but those in the very lowest classes had maids, and perhaps even nannies, who spent much more time with the children than the mothers. And although these people played very important roles in children's development, often being lifetime members of the family, they are not even mentioned in his psychoanalytic theories. We are only now beginning our recovery from this disaster of social engineering.

Big social projects are needed to offset the effects of the technological revolution. All technological progress destroys an existing social group and increases the space in which we operate. When the small town utility is absorbed in a large, regional utility, the group which ran it disbands and the area of service for that utility becomes much larger. Progress, per se, is not bad; but we must actively deal with the destruction of existing social groups by seeing to it that we make possible the creation of new social groups. In the past, a social group primarily existed in a given space. But with technological progress, the ever-increasing spaces in which we move may no longer provide workable boundaries for groups and, therefore, social groupings of the future may have to be time- or goal-based. That is, young people may dedicate several years to the service of their country in groups, or others may work hard together on a particular project for a short period of time. In the past, most social groupings were based on a geographical proximity. Now we need new groups which operate in larger spheres, held together by a common task, some of them time-limited. My father felt that five to seven years was the optimum span for any fixed social constellation in the arena of work. We must learn about the laws of other types of groups. Periods of shared service, as in war, can produce intense and lasting comradeship. Large national organiations provide lifetime friendships for others.

The macro-engineering of buildings and roads
is impressive, but easier than the macro-engineering applied to human society. Here follow some
dilemmas of our current human engineering: We
praise the family, and yet have a welfare system
with many regulations that tend to destroy families.
We have unemployment, but will spend a quarter of
a million dollars to train a handicapped person to
take a job away from an able person. We rehabili-
tate a chronic mental patient to the point where he
can hold a job, but as soon as he achieves this
economic success, we remove his medical benefits
and social support system without which he cannot
function. He can seldom earn enough to finance
his required medical and social support system. We
moan about high medical costs, but freely pay for
expensive, unproven treatments and will not pay for
much cheaper forms of treatment.

Finally, we must address the lynchpin in
successful macro-engineering of any kind--the problem of responsibility. Our moral system defines
individual sins and responsibilities, but in this
era of macro-engineering, individual sin is not
very important compared to collective and corporate sinning. For example, a large group of moral
people jointly might do something immoral and no
single person can be held responsible. The most
crucial immoral acts today are committed by large
organizations. Who will do the penance? Today we
sin not as individuals but as members of large
groups. We have no mechanism yet for a "macro-
morality" to go with our big projects. This we
must soon address, since we cannot afford to turn
our backs on it. Macro-engineering projects have
moral aspects. Their neglect has serious negative
social consequences. How can a corporation, an
artificial person, judged on a short-term financial
balance sheet, be moral? Some recent court
decisions are increasing the liabilities and
responsibilities of corporate directors. Perhaps
the legal concept of a corporation needs to be
revised to something more akin to a partnership
with greater personal responsibility for the
consequences of corporate action.

Frank Davidson, our Chairman, and Dr. Cynthia
Harris, a noted psychiatrist, both devoted a part
of their lives to making the concept of voluntary
service a reality. Their undertaking was

terminated by the advent of World War II. That
termination did not negate the validity of the
attempt, but delayed its execution. They began a
social macro-engineering project on a small scale,
as is proper. This country seems to be recon-
sidering the value of such efforts. A recent
report by the Department of Defense comes to the
conclusion that a National Service Program as
suggested earlier may be the only method for
fairly supplying the necessary manpower for a
successful military organization. Perhaps it is
time now to think about beginning again. A proper
perspective on our priorities, and the application
of sound macro-engineering to our social problems
could provide us a safe and better future.

Bibliography

Hausman, William, and Rioch, David; "Military
   Psychiatry: A Prototype of Social and Preventive
   Psychiatry in the United States." Arch. of Gen.
   Psychiatry, Vol. 16, June 1967, p.727

Huessy, Hans R. Caring, the Stepchild of Our
   Medical Care System. In Press.

James, William, "Moral Equivalent of War."
   Memories and Studies, New York, Longman's Green &
   Co., 1911

Preiss, Jack J. Camp William James. ARGO Books,
   Norwich, VT 1978.

Rosenstock-Huessy, Eugen, Planetary Service. ARGO
   Books, Norwich, VT. 1978

Rosenstock-Huessy, Eugen, Soziologie, Kohlhammer,
   Stuttgart, Germany. 1956 and 1958.

Dept. of Defense, "America's Volunteers", OASD,
   (MRA&L) Rm. 3E-773, Pentagon, Wash., DC 20301.

Overview

For those who prefer a long-range point of view, a "macro-chronic" timespan, the effect of technological efforts on future generations is a critical element of assessment. We need a macro-decision process which allows us to consider the desirability of macro-projects as well as their viability. Several contributors to this volume have addressed a number of the questions which arise in connection with the design and testing of such a macro-decision process. The development of a macro-decision system is itself a "macro" project which could have a notable impact on society. Large-scale enterprises have as much potential for benefit as for harm: it is up to us--to the public and the professions--to develop the means of selecting, using and controlling them in our best interests.

8. How Big and Still Beautiful? A Reprise

"Beauty is truth, truth beauty, --
                    that is all
Ye know on earth, and all ye need to know."
        --John Keats: Ode on a Grecian Urn

...And on the pedestal these words appear:
"My name is Ozymandias, king of kings:
Look on my works, ye Mighty, and despair!"
Nothing beside remains. Round the decay
Of that colossal wreck, boundless and bare
The lone and level sands stretch far away.
        Percy Bysshe Shelley: Ozymandias

In a bygone age, Keats found beauty, truth,
and immortality in the enduring work of a long-
dead and forgotten craftsman; Shelley found irony
in the ostentatious self-puffery of an ancient
macro-engineer. Today we ask again the poets'
questions: what is beauty, that we may know it
and create it; what is worth, that we may leave it
to posterity? Are we potters and artisans? Or
are we Ozymandias? How can we know?

To those who have adopted a long-range point
of view -- call it a macro-chronic timespan, extending beyond their own lifetimes -- the importance of these questions is manifest. Today's
technologies are as capable of producing destruction as they are hopeful of producing good works.
What is good in the short run may be disastrous
in the long. To achieve the admiration of the
second, fifth, or tenth generation down the line
may require sacrifice in the present. To avoid the

helpless curses of a posterity without options, we may have to forego major projects, programs or achievements in the present.

To those of the macro-chronic persuasion, these thoughts give rise to two operational questions:

1)   How can we measure the long-range effects of currently-proposed large-scale activities?

2)   If measurement were possible, how could we control or stop the activities thought damaging to the future?

Not all of the thinkers represented here -- and certainly not all of our engineers and voters and financiers -- are believers in the macro-chronic point of view. As a species, we are only a Darwinian eye-blink away from those common ancestors whose critical time-frame was measured in the hours required to procure food, or the days required to find shelter from the onset of winter. It cannot be surprising that most of us still think of project costs in terms of what we pay from our own pockets, and program benefits in terms of our own livelihoods. Especially in the industrialized West, we are a micro-chronic society, whose interest and attention are confined to its own life span.

To the micro-chronic, the poets' questions are meaningless. Beauty is what pleases me now, and worth is wealth, they say.

Ozymandias, as a micro-chronic thinker, probably died happy.

The Problem

George Kozmetsky, Frank Davidson, William Jones and James McHale have outlined the problem. In its applications of large-scale technology, says Jones, society is operating outside the limits of its own control systems. Although some major projects are easily identified, so that long-range impacts and the need for control can at least be considered, other activities of equal potential for harm are not seen as macro-systematic changes until

long after the fact. Macro-systems are capable of
self-generated growth, without consideration of
the dislocations they may cause in the future.

Kozmetsky points out that the managerial
planning and control model which makes large-scale
projects and programs feasible is not appropriate
for their evaluation. We have the ability to
organize,finance, and support huge undertakings,
but the same training which gives us that ability
prevents us from seeing the activities holistic-
ally. Traditional managerial models obscure the
purposes, interactions, and long-range dynamics
of the activities they have made possible. Macro-
programs become institutions in themselves, but
without built-in control mechanisms for self-
examination, evaluation, or alteration of their
program goals. We simply have no models to use in
evaluating macro-programs.

Davidson points out the essential role of
purpose in the structure and behavior of systems:
without purpose, there is no system, and without
system there is no possibility of control. A
fundamental prerequisite for evaluation is the
ability to "bring things together", so that purpose may be inferred, made explicit, and exposed
to public debate.

The world macro-problem has been aptly defined
by Willis Harman, cited here by Kozmetsky:

> ...perfectly reasonable micro-decisions
> currently are adding up to largely un-
> satisfactory macro-decisions.

McHale's paper provides a shocking case
example of the world macro-problem in action.
"Agribusiness" -- the collective noun referring to
all large-scale, capital-intensive agriculture --
is a macro-system whose purpose appears to be
maximization of current return on capital in food
production, through concentration of landholdings,
mechanization, and fertilization. The system is
driven by the rational decisions of well-trained
corporate executives, and its first-order effect
is the production of high yields per acre under
cultivation. Its longer-range effects, seen with a
macro-chronic viewpoint, are the demise of the
family farm and the exhaustion of agriculture's

most basic resource -- the land itself. Not noted
by McHale, but pertinent where "agribusiness" is
applied in formerly peasant-farming economies, is
large-scale unemployment and migration of the
peasants to urban ghettos, when farming shifts from
labor to capital intensity.

In agribusiness, farm equipment costing hundreds of thousands of dollars is operated by technicians, not by farmers. Economics dictates that
the equipment keep running; stopping it in order to
examine the soil is simply not justified. That the
land is becoming a sterile medium in which chemical
fertilizers may be converted to crops is not important to the macro-farmer or his industrial
suppliers; it is rational for them to assume a constant supply of the fertilizers on which they are
more and more dependent. The sum of their rational
decisions -- all based on the traditional management planning and control model underlying corporate operations -- could produce a macro-problem for
all food consumers, should the supply of fertilizers fail. But the managers of agribusiness have
not the models, the time-frame, or the language
with which to hear McHale's warnings.

The Impact of Immensity

Davidson, Cynthia Harris, and Ramon Barquín
have explored, from varied points of view, our
human reactions in the face of immensity. Here we
all have a common starting point, in our own
feelings as we experience the macro-structures and
phenomena of nature. The Grand Canyon, the
Himalayas, Niagara Falls -- all fill their wit-
nesses with feelings of awe, described by Harris
as a mixture of safety and security on the one
hand, with danger and excitement on the other.
Natural phenomena, too, produce these feelings, as
attested by those who have lived through hurricanes
or seen the coal-black disc of the sun in total
eclipse.

To Harris's ingredients of awe, I should add
another -- the feeling of individual insignificance
in comparison with Nature's scale and power.
Erwin Schell, founder of the department which became MIT's Sloan School of Management, summed up
that feeling with a story of a terrified three-

year-old's first visit to an ocean beach. When his
baffled parents finally quieted the boy and asked
why he was afraid, he sobbed, "Oh, Mommy -- ocean
so big. Billy so little." We are all puny and
powerless, beside nature's monuments. We, and
implicitly all of mankind, are simply --
insignificant.

Harris notes that the emotion of fear is an
ingredient of awe. She says, further, that some
people may be exhilarated and enhanced by their
confrontations with immensity, while others may be
overwhelmed. I suggest that the fear might be
directly related to the individual's own sense of
values -- that an ethno-centric orientation, seeing
the individual, the species, and Nature as parts
of one grand scheme, might contribute to exhilaration, while an ego-centric orientation, seeing the
self as the center and purpose of life, might lead
to feelings of some inadequacy and meaninglessness
when confronted with the scale of natural grandeur.

The ethno-centric value system would also be
consistent with a macro-chronic time frame, and the
ego-centric with the micro-chronic. And, as value
systems are generally thought to be culture-depen-
dent, we might expect a difference in reaction to
immensity in peoples of differing origins and backgrounds. It is difficult to imagine, for example,
a member of the Hopi Indian tribe being overwhelmed
by the spectacle of the Grand Canyon. To the
American Indian as to followers of many Eastern
philosophies, as I understand them, humanity and
nature are parts of an integral whole. They belong
together, and where there is a sense of unity and
eternity, there is no room for fear of the immense.

Human values are at the forefront of Barquín's
tour through the archaeological remains of ancient
macro-engineering projects, as well. He notes that
the values of their builders--particularly their
inferred dreams of immortality -- have determined
what has been done in the past. Further, the
accomplishments of the past have influenced the
values of today, and we may surmise today's projects will leave a mark on the generations to come.
For good or ill, what we undertake will build the
world in which posterity develops its own sense of
"good", "bad", and "big."

Barquín's speculation on the relationship between macro-engineering and the search for immortality is best illustrated, for me, in Harris's delightful story of Mies van der Rohe. The master architect, surveying his finished work, nods his head and muses, "Well, they will know we were here." It might have been Imhotep. The master builder leaves enduring evidence of the accomplishment, as proof that men were there. Who "they" are is immaterial. The designers and laborers of Hadrian's Wall, Stonehenge, the Tower of London, the Roman Forum, and the Rhineland castles had no notion of American tourists, but their works give me a sense of pride, awe, and continuity with the past. Whether the constructions were built for war or worship, commerce or comic relief, does not matter to me. They were built by people, and they remain significant. I am a person, and I share the builders' satisfaction in saying, "They will know we were here." There is meaning in being human.

Barquín, Kozmetsky, and Larry Meador have highlighted the essential role of organization and information handling in the accomplishment of large-scale activities, now as in the past. That organization itself is a macro-activity with its own set of deleterious side effects is recognized in the literatures of Organization Behaviour and Sociology but only rarely in the field of engineering. Let me follow Cynthia Harris's Niagara Falls anecdote with one of my own, to illustrate.

I recently had the chance to re-visit the Falls, with family and a visiting Polish graduate student. Arriving in Buffalo late on a Thursday night, we found our friends there extremely skeptical about the Polish guest's chances of admission to Canada, where the vista of Niagara is far more impressive than the view from the American side of the river. Even if we succeeded in getting Mira into Canada, they said, we would find it impossible for her to return to the United States; her visa was clearly marked "one entry only," and the U.S. Immigration and Naturalization Service was known to interpret that phrase strictly. One Buffalo friend had recently spent three days trying to bail out a visiting Japanese businessman who had casually walked into Canada to see the Falls; that visitor had to travel to Toronto to procure a second American visa, for re-entry to the country.

Hopefully, I took Mira to the Canadian consulate in Buffalo. After a long wait in line, we learned it takes two full working days to procure a visa for a Polish passport: Mira's could be ready by the following Thursday morning. Since she would be back in Gdynia on the following Thursday, it would be impossible for her to see the Falls from the Canadian side. There was nobody in the office who could talk to her directly, on the spot. There was no superior officer we could talk to directly, or reach by telephone. There was no channel for handling exceptions. No, nothing could be done. A call to the U.S. Border Patrol indicated nothing could be done about the problem of re-entry on a one-time visa, either. Period.

If Mira is typical, the Poles are a very persistent people, who take their bureaucracy with a grain of Baltic salt. Getting into Canada became a challenge. Mira spent hours practicing answers to the possible questions border guards might ask, trying to assume a harmless American accent. No luck. She practiced crying, just in case. On Saturday morning, we set out for the bridge.

Stopping first at the American side to get a feel for the re-entry problem, we found an officer who listened sympathetically. "See me on your way back, and don't let anybody take away this slip of paper stapled to your passport," he said. We walked on to the Canadian entry point. "It can't be done," said the guard, and, "highly irregular," said his boss. But a half-hour later, Mira had an official Canadian visa.

The Falls are immense, and we spent the whole day getting to know them, from all possible angles and elevations. Less thrilling, but impressive in its own right, is the Niagara Power Project -- the gargantuan but unobtrusive macro-engineering project which provides electricity to the Northeast while respecting the grandeur of the Falls. We all came away from the day with a dual impression -- the sense of awe and wonder at the Falls, and the sense of conquest. Nobody conquers the Falls, of course-- not even the barrel-riders -- but we had beaten the organizational system which surrounds them. To Mira, that demonstration that the individual can defeat an impersonal bureaucracy was just as important a thrill as the sight of the torrent itself.

Criteria for Judging Big Things "Good"

Several of our writers here have enumerated criteria by which proposed macro-activities might be evaluated. Jones suggested measuring the demands of the activity on our natural resource base, the number of people involved in its construction and use, and the timespan in which its side effects would require safeguards. Harris suggests that the purpose of the activity must be worthy, and appropriate to the scale of resource use, and further that the activity have the property of "belonging" to all of us; Davidson also stresses the careful examination of purpose. Barquín adds the need for specific examination of the activity's impact on human values.

Several writers have also referred to the criteria for defining something as "big", without specific attention to its goodness or badness. Among these criteria are sheer bulk; complexity in relation to what has gone before; technological challenge; cost, relative to the capacity of the society contemplating the project; duration of the design and construction; duration of the activity's utility, and the cost of removing it if it should become annoying or dangerous to the citizens of the future.

Implicit in all these thoughts is the notion of control. We are technologically capable of destroying the future. We need criteria of "bigness" in order to tell what to control, and criteria of "goodness" to see how to control. Many speakers have suggested that in our society macro-projects and programs are possible only with government support, and therefore, they are easy to isolate. Jones disagrees, noting the incremental nature of some developments whose impact is monumental. McHale would join Jones: agribusiness is a perfect example. Harman's problem statement, that the grand sum of perfectly rational micro-decisions can produce a macro-problem, remains as our dilemma.

The evaluative criteria listed appear to fall
into two broad classes. Utility, economy, and
practicality seem to cluster together to produce an
overall criterion of project or program viability.
Safety, audacity, immortality, and humanity seem
to add up to an overall criterion of its
desirability. A macro-activity may possess either
attribute without the other; only if it possesses
both should it go forward.

And this is the heart of the dilemma. The
management planning and control model used by
business and government leaders is a magnificent
machine for evaluating the viability of a proposal.
We have no comparable machine for evaluating its
desirability. Without such a mechanism, all our
talk of criteria for judging "goodness" is a waste
of time.

The Macro-Decision Process

The management planning and control model does
not simply ignore the desirability of a proposal.
Rather, it literally devalues the long-range future
where most of the factors making up desirability
(as defined here) are focussed. The devaluation is
accomplished through the use of the most basic
tools of financial management -- cost/benefit calculations and discounted cash flow analysis (DCF).

Discounted Cash Flow is simply the process of
comparing what you will get from a dollar's investment in the project, against what you would
gain from putting the same dollar in a savings
account. The technique asks, "What is the value,
right now, of the returns we will get from this
investment in, say 25 years?" It turns out that
the present value of a dollar that far away in the
future is very small, at any realistic interest
rate. Effectively, the long-range future is discounted out of existence. Only the short-run
returns of the possible investment count as benefits to the financier. And managerial vision is
constrained to the time horizon of its tools.

In teaching the skills of strategic decision-
making to the managers of the future, instructors
of Business Policy focus on four major questions
which must be considered in the formulation of
strategy:

1) What <u>might</u> we do? (What alternatives have we?)

2) What <u>can</u> we do? (What resources, strengths, and capabilities can we bring to bear on the problem?)

3) What do we <u>want</u> to do? (What is consistent with the values and desires of the organization's owners and managers?)

4) What <u>should</u> we do? (What are our obligations and responsibilities to society as a whole?)[1]

The same four questions are applied to the evaluation of all macro-proposals. The problem we face here is that the questions are asked and answered by the same people -- by those with the ideas, the resources, and the power to decide. The first question -- what we might do -- is answered by those individuals of creative genius who come up with the ideas. They are the visionaries, the dreamers who say, "This is possible."

The second question -- what can we do -- is answered by the financiers and managers (in business and/or in government) who can command the necessary resources. They are the technologists, the organizers, and the movers who say, "This is viable."

And, in our current macro-decision system, that is all that is required to get a project started. The third and fourth questions -- adding up to what is desirable -- are answered by the same people and institutions which generated the proposal and the resources in the first place. And their answers are biased. Of course, the dreamer wants to see his dream realized. Of course, the sponsoring institution wants to proceed. And of course, they assume that the project is desirable to the whole of society; it is obviously desirable to <u>them</u>.

---

[1]Christensen, C. Roland, Kenneth Andrews, and Joe Bower. <u>Business Policy: Text and Cases, Fourth Edition</u>. Homewood, IL: Richard Irwin, 1978.

Our current macro-decision model contains no mechanism save serendipity for testing the value judgments of the self-interested institutions which promote the projects. All of us involved in the writing of this volume have based our contributions on our own value systems and our own senses of social responsibility. Whose values should apply to the evaluation of macro-proposals?

Certainly not my own, at least not in isolation. Some of the proposals contained in the second half of this book are absolutely appalling to me. If they were to gain financial support, I should feel constrained to take to the streets, if necessary, to fight them. Better there were some mechanism to resist those projects <u>before</u> they were judged "viable" by some decision-maker, because once that judgment is reached, huge resources are suddenly put to work to convince the public of the project's desirability. By the time the juggernaut begins to move, its mass is enough to crush dissent.

Harris points out that the essence of freedom is the ability to say "no" to challenge. McHale's experience indicates simply saying no is not enough; the people riding the juggernaut are not obligated to listen, and they have a loudspeaker system capable of drowning out the lone negative voice.

Alternatives for a New Macro-Decision System

A few of the writers here have suggested possible mechanisms for wider involvement of all of us in answering the questions about the desirability of proceeding with identified macro-activities. Kozmetsky believes the matter may be safely delegated to established business leaders. Davidson feels the education and training of project managers may hold a key for better macro-decision-making. Harris suggests deliberate allocation of project resources for public education might promote a general feeling of "ownership." Huessy advocates an organized application of social psychology to all macro-proposals, and Kozmetsky the development of a field of social engineering to match the hardware engineering we are now capable of.

It is premature to attempt evaluation of these ideas now. They are young and vulnerable, and deserve the chance to mature before undergoing criticism. Further, they are hardly an exhaustive list of "what we might do" to provide a new, more effective, more efficient, more broadly-based macro-decision system. A possible next step in designing such a system might be turning our attention to development of a comprehensive set of alternative decision-system designs. Given such a set of specific plans, we could subject them to evaluation according to the criteria we have developed here, combine or modify them to remedy weaknesses, and gradually converge on a system design which is both desirable and viable.

In our deliberations here, we have agreed that some form of macro-decision system is highly desirable. Such a system would itself be a macro-activity of tremendous impact on society. Building upon the values of today's populations, it would shape the values of tomorrow's. As those values are applied, they should move us toward new and more constructive applications of large-scale technology. They should point us toward calm consideration of the inherent trade-offs in any macro-proposal -- trade-offs between the present and the future, between labor- and capital-intensity, and between large and small scale.

In our society, bigness is a fact of life. It will not and cannot simply go away because it makes some of us uncomfortable. Bigness is not intrinsically bad; it has as much potential for benevolence as for destruction. How we use it will make the difference, and how we control it will govern how we use it. The challenge facing all of us is the design of a macro-control decision system -- one which considers what "we" want to do and should do, as well as what we might and can do. It must be possible to make the big, beautiful.

Part 2

Cases in Point: Land, Sea and Aerospace

Transforming the Urban and Rural Environments

Aerial view of the proposed new capital city of
Nigeria. The central axis of the city traverses
four low, rounded hills, interspersed with minor
stream valleys. Important public buildings will
be built where the axis crosses the high point of
each hill.

Overview

The proposed new capital city of Nigeria has been described as the largest free-standing new town project ever undertaken, with an estimated population of 1.6 million people by the year 2000 and a probable area of 25,000 hectares. Construction is scheduled to begin in 1979, with development to proceed in stages over a 20-year period. The master plan for the new city emphasizes adequate transportation, housing, balanced employment and public service infra-structure, and the use of technologies appropriate to the culture and management capabilities of the country. Many issues have not been resolved, but in view of the commitment of the present government and the degree of popular support, Andrew C. Lemer guesses that the planned capital city of Nigeria, located in the interior in an ethnically neutral area, will indeed become a reality.

9. Macro-Project in West Africa: Planning Nigeria's New Capital

After ten years of military rule following the Civil War, Nigeria will soon return to civilian government under a new American-style constitution. The nation's buoyant economy, running on petroleum, is supporting major expansion of new infrastructure and rapid industrial growth. Government planners are facing difficult problems due to accelerating urbanization and extreme poverty.

In this context, Nigeria has decided to embark on construction of a new Federal Capital City (FCC). A consortium of American consulting firms* was selected, through international competition, to plan and initiate design of this major macro-engineering project.

The experience of building a city of this scale is still in the future. We can only discuss the basis and progress of overall master planning which is nearing completion, and make conjectures about lessons to be learned.

Background

Construction of the new Federal capital is scheduled to begin in early 1979, with the principal functions of Federal government to be moved to the FCC by 1986. The city is planned to grow to a population of 1.6 million people by the year 2000,

---

*International Planning Associates is a consortium of Archisystems, Inc. (a division of the Summa Corp.), Planning Research Corp., and Wallace, McHarg, Roberts and Todd.

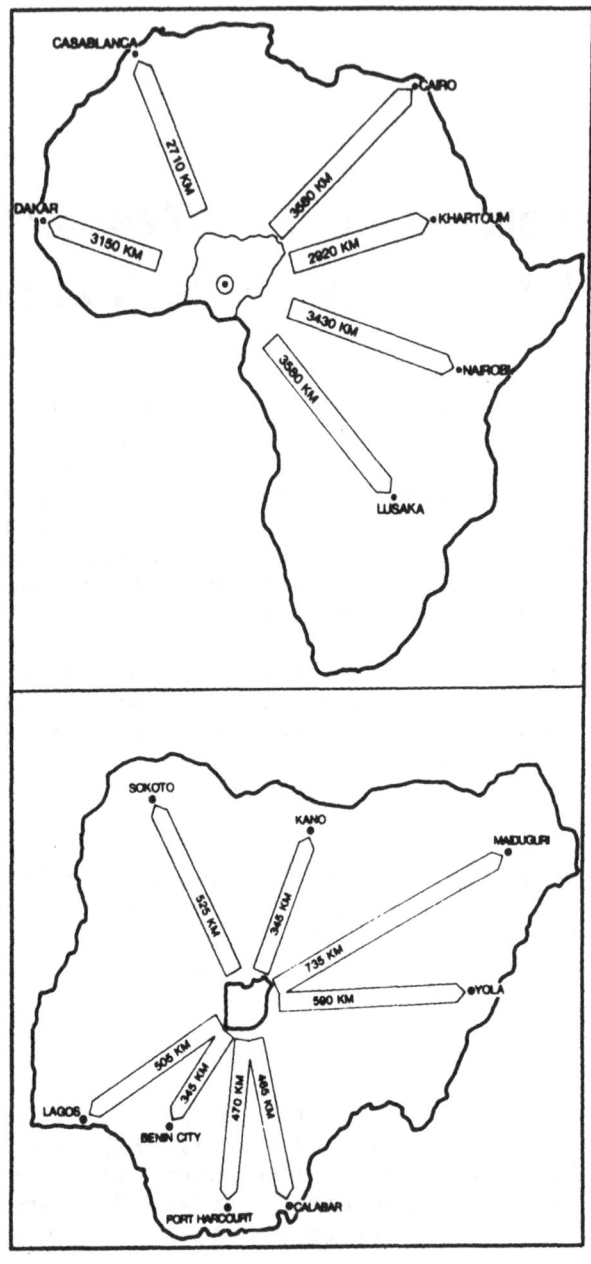

Figure 1. The new Federal Capital
City of Nigeria is centrally located
to points within Nigeria as well as
to the rest of the continent of Africa.

and to occupy an area of approximately 25,000 hectares. The site of the FCC was selected to accomo-
date an ultimate population of up to 3 million
people.

The reasons for moving the capital from Lagos,
Africa's largest city, are several: to locate
government in a centrally positioned and ethnically
neutral area, to escape the chaos of a city which
has long-since outgrown the capacity of its infrastructure, and to provide a showcase for national
unity and modern African urban development.

At present, the city has no name. While no
formal requests have been made or contests run to
find a suitable title for Nigeria's new capital,
active public interest has been reflected in the
proposals for names, which frequently appear in
letters to the editors of the nation's principal
newspapers.

As with any project of this magnitude, there
are serious questions regarding the viability and
relevance of the new capital. With the problems of
poverty and low quality of life widespread, can the
nation afford to concentrate a significant portion
of its resources in a single location? At a time
when the development of new cities is being ques-
tioned in many developed countries, as a reasonable
strategy for managing urban growth, can the FCC be
expected to meet the ambitious objectives set for
it? With the uncertainties of the international
petroleum economy and a dynamic domestic political
situation, can Nigeria maintain the long-term
commitments required to carry the project to a
successful completion?

Location of the Capital

The FCC is located within a centrally
positioned Federal Capital Territory (FCT). The
FCT is under direct administrative control of the
Federal Capital Development Authority (FCDA), a
specially appointed agency of the Nigerian government, whose chairman holds ministerial rank.

The FCT is situated just north of the humid
lowlands of the Niger/Benue Rivers trough, but
south of the driest Sahelian portions of Nigeria,
in an area called the Middle Belt (Figure 1).

The climate is tropical, and exhibits distinct wet and dry seasons. Monthly precipitation is estimated to range from negligible in the November-January period to a peak of 300 mm.

The FCT is generally characterized as a tilted plain rising slowly in elevation from southwest to northeast. Several ranges of low mountains and numerous rocky knobs rise above this plain. These rocky knobs are particularly prominent in the northeastern portions of the FCT, providing a strong visual image and distinctive landmarks for the site of the new capital. These landmarks play a strong role in the proposed FCC plans, providing a setting of appropriate variety, interest, and image for Nigeria's new capital.

The central location of the FCT and the topography of the FCC combine to produce a significant variation in climate as one goes from southwest to northeast parts of the territory. The rainy season is shorter and temperatures somewhat lower in the northeastern areas. Of the several areas of adequate buildable land within the FCT, the proposed site of the FCC offers the best climate.

The FCC Plan -- An Overview

The master plan of the FCC has been developed to be responsive to a wide range of factors:

- The plan to have a single contiguous urban mass is a result of regional land suitability, micro-climatic factors, and a desire to minimize infrastructure costs.

- The decision to have a dominant center rather than multiple centers is based upon a desire to create an appropriate image of importance for the seat of government and to support efficient functioning of government.

- The location of the center was chosen to use prominent natural features of the site as a means of emphasizing the uniqueness of the FCC.

- The curvilinear form of the plan is based upon topography and environmental factors,

combined with a desire to maximize transport efficiency.

- The structure of residential sectors and
  service systems is based upon existing
  characteristics of Nigerian society, combined with the nation's objectives for its
  future development.

The most important single factor in determining the physical organization of the FCC is
transportation structure. Two parallel transit
spines are brought together at the central core,
which is the main government, institutional, and
commercial complex. Bounded by peripheral highways, the FCC follows these corridors, assuming a
form which is essentially a "wedges and corridors"
radial plan deformed to a linear shape. The great
majority of residents will be within walking distance of the transit system.

Along the two corridors, residential areas are
organized into distinct sectors. Each sector, with
a population of between 75,000 and 200,000 people,
is intended to be a mini-city providing its residents with local employment opportunities, community and commercial services. Each sector is further divided into districts, their sizes determined by population necessary to support schools,
health care facilities, markets, etc.

Separating the corridors and surrounding much
of the city are reserved urban parks and open
space. Stream valleys comprise an important portion of this reserved open space, providing natural
drainage, and aesthetic benefits.

Industrial development, which we estimate to
be a relatively small element in this government/
services-based city, is concentrated in estates
located at lower elevations than the residential
and commercial areas. Intersections of the road
network provide these areas with good accessibility.
An international airport is located further to the
west and linked to the city center by a limited
access expressway.

Housing

As one of the most basic of human needs and

the most direct point of contact between the FCC
and its residents, housing is a major concern in
the FCC. The housing program attempts to strike a
balance in the conflict between the standards which
the government wishes to provide its citizens and
the costs which the FCC's residents and the government can afford.

As with the use of modular sectors to provide
flexibility in the overall plan, the housing program is intended to respond to a variety of alternative institutional and financial approaches to
implementation. Land is reserved in the plan to
permit expansion of the numbers of housing units
and sizes of individual units -- as incomes rise --
without having resultant uncontrolled sprawl of the
FCC.

Within the individual housing units, it is
assumed that finished work will be reduced
and self-help -- "sweat equity"-- will be employed
for upgrading. However, a detailed housing design
has not yet been undertaken, and it is likely that
the FCC's housing program will undergo review and
revision as development proceeds.

Public Services

After basic shelter and adequate and balanced
employment, provision of access to good quality
public services is a major aim of planning. While
actual quality of service delivery is dependent
upon the organizations having final responsibility
for operations, service facilities have been sized
and located for accessibility.

Schools have been used as foci for residential
organization and community activities. Implementation of universal primary education decrees and
assumed increases in enrollment rates at all levels
make education a major land use within the FCC.

Health services are to be provided in an integrated system, from a major hospital complex with
teaching facilities down through specialty clinics
to frequent local clinics and baby care facilities.
Libraries, post and telegraph facilities, police
and fire services, and cinemas, theaters, and other
local cultural and recreational facilities are dis-

tributed in accordance with Nigerian practice and recommended international standards.

As the center of the nation, the FCC will include a number of unique cultural, recreational, and governmental facilities, many of monumental significance. It is expected that these facilities, including a national stadium, library, capital and executive residence, will be designed by Nigerian architects. Consideration is being given to holding an invitational competition for architectural design for the central area of the city.

Implementation and Incremental Growth

Development of the FCC is planned to occur in stages over a 20-year period. Staging is based upon the residential sectors, allowing flexibility to speed or slow the process without upsetting the functional integrity and viability of the FCC as a whole. Population growth, and thus plans for implementation, are based upon projections of Federal government employment and policy toward decentralization.

Imbedded in the population projection of 1.6 million people by the year 2000 and related to a city size estimate of 25,000 hectares are a large number of assumptions regarding labor force participation, family size, relative incomes, and employment characteristics. An important factor in the FCC's growth will be the "informal" sector of its economy, the artisans, vendors, domestic servants, part-time workers, etc., who comprise a major portion of the work-force in many developing nations.

A close partnership of public and private sectors of the Nigerian economy will be required in the FCC's development. While the new city will be built to serve the functions of government, a large proportion of the FCC's costs will be associated with activities of non-governmental business and private individuals.

The plan reflects a principle that the proper role of government in developing the FCC is to provide necessary infrastructure, social overhead, and economic base to support natural growth and

development. Much of the investment which will
occur in the course of this natural growth will be
made by the private sector.

Project Management

In general, a "fast track" approach is required for the FCC in planning and design as well
as in construction. Construction will begin on the
initial residential district while many of the
large-scale planning issues are still being
resolved.

The Federal Capital Development Authority
(FCDA) was created to bring the concept of a new
capital to actuality. This ministerial level body
has full governmental authority to develop and
administer the FCC, and is limited only by its own
technical and managerial capabilities, and by yet
to be established Federal budget constraints. The
FCDA is currently undergoing rapid increases of
personnel and is faced with difficult decisions
regarding basic approach to continuing project
development and construction management.

Logistics

While the city for 1.6 million people will be
a major challenge to complete, we feel that the
most difficult goal is that of moving government
by 1986. It is planned to have a population of
150,000 people by that time, housed in permanent
facilities. The initial infrastructure and
housing required will represent a heavy demand
on the construction sector. Peak demands for
materials, equipment, labor, and transport services
are projected to occur during the 1982-84 period.

Through 1981, we estimate that the existing
system for supply of construction materials will be
adequate to avoid major shortages and inflationary
pressures. However, most of these materials are
imported, and as FCC material demands increase, we
feel that some mechanism for advanced procurement
and stockpiling will be necessary.

Similarly, the initial workforce is estimated
to be available, and might be housed in temporary
quarters in nearby villages. However, as the labor
force grows to its estimated peak of 44,000, hous-

ing will be required on-site. We anticipate that initial construction will be permanent housing units, which may be used for this purpose.

While it is estimated that the existing transport system has adequate capacity to meet demands during this period, dedicated port facilities and coordination of trucking are essential if the delays endemic to Lagos are to be avoided. Over the longer term, barging of materials up the Niger River appears to be a most promising means of supplying the FCC.

Is a New City a Good Idea?

It is estimated that the urban population of tropical Africa is growing at a rate in excess of 5 percent per year, significantly faster than the overall population growth rate of 3 percent per year. The impact of the coming "astonishing population increase" will be felt mainly in existing cities.[1]

Lagos, the largest city in black Africa, is an early and graphic case in point. The population of Greater Lagos, approaching 3.5 million people, will have increased by more than the 75 percent projected for the 1970-1980 period. The nation as a whole will have grown approximately 32 percent.

This growth has not been accompanied by comparable expansion of urban infrastructure and services. Lagos is "well on its way to becoming the Calcutta of Africa."[2]

Yet, in spite of such massive urban growth and expectations for its continuation, the population of tropical Africa is still overwhelmingly rural. It is estimated that only 11 percent of the approximately 242 million people in the 35 countries of the region lived in settlements of 20,000 or more individuals.

---

[1]International Planning Association. Draft Master Plan for the Federal Capital. Washington and Lagos, September 1978.

[2]Rosser, C. Urbanization in Tropical Africa: A Demographic Introduction, International Urbanization Survey, New York, Ford Foundation, 1972.

Appeal of a New City

Hence, many of the people coming to Lagos and the other major cities of the region are likely to be experiencing their first contact with any form of urban life. Lagos, with its sprawling slums, erratic water and electricity is a clearly ineffective mechanism for productive integration of these migrants into the urban economy.

The concept of a new city is thus especially appealing. The FCC is a potential magnet to attract migration from Lagos, while providing relatively immediate relief for government operations severely hampered by the inefficiencies of Lagos' congestion.

But even if one accepts the desirability of a new city, the size is still a major issue. In spite of extensive research, no clear-cut relation between urban sectoral efficiency and city size has been proven. It has, however, been argued that larger cities are required to support large services sectors, particularly the high-order "informal" sector typical of many LDC cities.[3]

It is estimated that in the year 2000 more than 40 percent of the total employment of FCC will be in this informal sector. Overall labor force participation is estimated at 52 percent of the population.

If the experience of Lagos -- and to a somewhat lesser extent, other cities in Nigeria -- is applicable, it is unavoidable that the FCC will grow large. Indeed, it is hoped that the FCC will shift migration and urbanization more toward the central areas of the nation, into the "Middle Belt" where development has lagged behind both north and south.

The challenge will be to control the early years of the FCC's growth. Migration to the job opportunities of future construction will have to be restricted to assure that population does not expand beyond the capacity of infrastructure.

---

[3]Richardson, H.W. City Size and National Spatial Strategies in Developing Countries, Staff Working Paper 252. Washington: World Bank, 1977.

Balanced Planning and Appropriate Technology

A broad view must be taken of what infrastructure is required to support population growth.
The public services of the FCC are as important as
the physical systems of transport, water supply,
and sewerage. Balanced development of these social
systems represents an early investment in human
resources. This investment will be extended to
education. A program of manpower training is
planned to provide construction labor for the
city's development.

Throughout planning of both physical and
social systems of the FCC, "appropriate technology"
has been a matter of continuing concern. While a
new city provides an outstanding opportunity to
apply new ideas and emerging technologies, lack
of technical sophistication in the society as a
whole dramatically increases the risks of total
failure of these new ways. This lack of sophistication is manifested in the absence of personnel
trained to operate and maintain advanced systems
and inexperience of the population in the use of
such systems.

If the FCC can demonstrate the value and
applicability of new technologies for the West
African city, it will have served a valuable function, one that is unlikely to be possible with
existing cities. However, a key to success will be
judicious blending of new technology with old
traditions. For example, the transit spines which
are the major structural feature of the FCC are
planned to be busways -- paved roadways with use
restricted to public transit vehicles. Such
restricted-use roadways are new to Nigeria, and it
is not unlikely that stringent measures will be
required initially to assure that access to the
roadway is indeed limited.

For example, modern expressways in Lagos are
often obstructed by cattle crossing the roadway.
Barriers have been constructed in the median of
some roads in an unsuccessful attempt to encourage
people to cross at available pedestrian bridges or
underpasses rather than crossing active traffic
lanes.

The vehicles using the busways will not necessarily be only large transit buses. "Mammy wagons"

Figure 2. Artist's concept of the FCC's parliamentary buildings from the Mall.

and other traditional forms of transit, which comprise an extremely important part of the urban
transport system in Nigeria, should play a major
role in the FCC.

In contrast to the transportation system,
advanced technology is planned as a means of minimizing maintenance and operating difficulties in
telecommunications. It is anticipated that modular
digital solid-state systems and fiber optic transmission may be employed. Centralization of maintenance and network management will then permit a
small group of trained technicians to care for the
FCC telecommunications.

Symbol for the Nation

As a potential symbol for Nigeria's future,
the new capital follows a long tradition. Canberra,
Brasilia, and Washington are national capitals
which have grown from grand plans placed on virgin
land.

As Eldredge states in his study of world
capitals,[4] the capital "bears the symbolic torch
for the entire nation -- the image the chosen city
creates is a powerful factor in national status and
internal self-conception." The FCC is sited and
planned to play this role.

Image is initially provided by the landscape.
The FCC's central area will be aligned to focus on
Aso Hill. Major symbolic architecture--monuments,
legislature, etc.--are to be located on low hills
along the axis (Figure 2).

Image will, of course, be provided primarily
by design of individual buildings and sub-areas
within the FCC. It is envisioned that there will
be an international competition for design of the
central area.

In addition to the visual image, the FCC will
be symbolic in the standards of urban life it is
intended to provide. Planning decisions have been

---

[4]Eldredge, H.W., ed. World Capitals: Toward Guided
Urbanization. Garden City, NJ: Anchor Press/
Doubleday, 1975.

made to try to maintain a balance between quality
and affordability, in an effort to assure that the
plan will be implementable. A new city avoids the
physical-psychological obstacles to implementation
which an existing urban structure may present.

The FCC as Feasible Macro-Project

Nigeria's Federal Capital will be the largest
free-standing new town project ever undertaken.
However, the problems which a project of this scale
might be expected to experience are in large
measure mitigated by the time period allotted to
its implementation.

As discussed previously, the most difficult
years will be the period between now and 1986. To
produce a functioning urban environment to house
the nation's capital and 150,000 residents requires
construction of quantities of infrastructure,
buildings, and monuments out of proportion to the
population as a part of the total 1.6 million
people scheduled for the year 2000.

Only preliminary estimates of costs and financing requirements have so far been made, and if the
lessons of other major projects are relevant, these
initial estimates are likely to be low. Yet even
allowing for cost growth, the Federal government's
share of annual investment in the FCC appears to
be well within the limits of the nation's current
budgets and practical capacity. Any future real
increases in oil revenues would assure that development of the FCC can be completed with no strain
on the economy.

There may be problems in the early years with
supply of labor and materials. Congestion of the
transport system and the desire to minimize impor-
tation of construction materials and skilled labor
could slow initial progress. As discussed previously, logistics planning has been undertaken, and
with careful management, these problems can be
minimized.

Indeed, it is development of adequate management capacity which may be the principal challenge
to successful development of the FCC. Estimates of
personnel requirements indicate that substantial
growth of FCDA procurement and contract adminis-

tration staff will be required if project management is to be provided "in-house."   There has
already been a major increase of architectural
and engineering personnel, recruited primarily from
several eastern European and Asian countries.

As with the problems of logistics, there are
a number of ways in which these potential management difficulties may be solved.   The apparent
commitment of the government to the FCC suggests
that such problems will, in fact, be solved.   Popular interest, as reflected in newspaper editorials
and experience of members of the planning team in
Nigeria, appears to indicate that this commitment
will survive the transition to civilian rule.   It
thus seems quite possible that the plan for a new
Federal Capital of Nigeria will indeed become a
reality.

Overview

One possible solution to the problems of the increase in scale of contemporary society would be to utilize the surface of the sea for living space and to make greater use of our waterways for marine mass transit. According to John P. Craven, the cultural advantages of water-based cities have been demonstrated in the neolithic lake societies, and the technological feasibility of ocean cities has been demonstrated by the numerous varieties of offshore platforms produced in recent years. The increased use of marine mass transit on rivers and across oceans would afford designers an opportunity to furnish examples of ideal urban characteristics. Now that technical feasibility has been demonstrated, it only remains for a favorable social-political climate to emerge for ocean cities and urban marine mass transit to be granted a fair trial.

10. Cities of the Future: The Maritime Dimension

An overriding and apparently inexorable feature of the history of societal organization is the
increase in scale. The size of cities, of buildings, of athletic stadiums, of theatres, of
thoroughfares, of homes, of shopping centers; the
volume which society allocates to each individual
at home, at work, in the office; the size and
velocity of vehicles for transport of passengers,
the number of passengers per vehicle; the volume of
communication, the rate of transmittal and the
distances over which they are transmitted and the
number of recipients of each communication; all
show monotonic increases. Only a few isolated
exceptions to this trend can be cited such as the
Caracalla baths of Rome, the 2,000-passenger ferry
boats, or the estates of the affluent. This growth
in scale is so apparent and in recent times so
accelerated that analysts have succumbed to the
temptation to represent this growth by empirical
equations of the form Ce^{at} and to thereby extrapo-
late a growth which would constitute a clear and
present danger.

In the face of this perception or mispercep-
tion, compelling arguments have been made that
"small is beautiful." Indeed, a number of small
groups of individuals who, failing to perceive
that they are, in fact, numbered among the world's
most affluent, have been able to afford the imita-
tion of the lifestyle of a previous generation when
the planet Earth had a much smaller population in
relation to its total resources.

If we recognize that e is a constant of mathematical convenience ($\int_0^1 e^x dx = e^x$) and not a constant of nature, that growth processes are in general non-linear, and that growth rates are very different for different functions of living, then we are ready to disregard dangers which are the product of over-simplification and ask more exciting and relevant questions. We should note that many of life's functions are not amenable to changes in scale and are not likely to change for many decades or even centuries. These include but are not limited to life expectancy, athletic capability, the rate of speaking, the rate of writing, the capacity of the brain to receive information in rate or total quantity, the period of gestation, the years of childhood and adolescence, the productive years, the average size of the family, of social groups, clubs and fraternal orders, etc. Although most of these factors have been extended or expanded, they represent nearly constant and size-limiting factors in the scale of society.

This paper, therefore, examines the more exciting question: How big can the other scales of society be in order to benefit from the economies of scale and still be beautiful? The word "beautiful" is used in the sense of aesthetically and sociologically satisfying to the populations of such societies.

We shall argue in this paper that at least one order of major increase in the scale of society (the size of cities, the population of earth, the size of vehicles of transport, etc.) can be accomplished while restoring or establishing a "beautiful" society. The argument depends for its validity on the utilization of the surface of the sea, the bays, the rivers, and tributaries of the world as usable space. The argument is equally dependent on the assertion that technological breakthroughs have occurred which will eliminate the two major inhibitors to living on the sea surface. These are the perils of the sea associated with storm and aggravated sea state, and motion sickness and psychological discomfort associated with sensible motion of the world in which we live.

Before we can make this argument, we must define what we mean by a "beautiful" world.

Although it is doubtful that agreement can or will be obtained for any such definition, an attempt has been made by this writer and the Japanese architect Kiyonori Kikutake to define the design requirements for an urban society and, more specifically, the city.

In broadest terms, an ideal city will: 1. meet minimum requirements for the mechanics of living, 2. provide protection against natural and man-made depredations or aversive stimuli, and 3. provide ample opportunities for each individual to satisfy his or her needs for reward, i.e., the provision of abundant opportunities for the widest range of positive stimuli. If we examine each of these factors in some detail, we may see those which are amenable to satisfaction by increases in the scale of society, and those which are not.

The minimum requirements for survival are well known and all city planners and designers are careful to include these factors in their layout and plan. They include: an economic supply of water, food, clothing, and shelter for all the inhabitants; non-polluting, rapid, comfortable, flexible, and immediate transportation for people who may be carrying from 40 to 80 pounds of goods; non-polluting and non-obstructive transportation of goods in quantities in excess of 80 pounds; telephone, radio, television, printed media, mail, accounting, and computational communications; non-polluting, efficient, and effortless disposal of wastes—human, solid, liquid, organic, toxic, and non-toxic.

These minimum requirements apply equally well to a prison community, an army, a ship's crew, space voyagers, or any other aggregation or colony of individuals who are destined to live out a normal span of life. To the extent that a city is a voluntary association of individuals, satisfaction of life-support criteria is not sufficient to insure its viability. The Thorndyke-Skinner model of human behavior tells us that people will repeat behavior and performance when that behavior (even though accidentally or randomly initiated) is rewarded. This model also tells us that although people will adapt to, ignore, or even find reward in mild irritants, they will react strongly to destructively punitive phenomena or aversive stimuli. The avoidance of aversive stimuli is thus

another mandatory criterion for city design. These
avoidances include: freedom from pollution--atmos-
pheric, water, land, noise, and time; protection
from fire, flood, tsunami, earthquake, hurricane,
tornado, volcano, and other natural disasters; military defense and protection against crime, civil
disorder, and civil disaster; protection against
sickness, disease, epidemics, and bodily injury;
provision of adequate public and private health and
hospital facilities.

Fulfillment of these design criteria will make
the city livable but not lovable. It is the abundance and availability of positive reinforcements
that give a city charm and desirability, that make
it attractive and induce people to live in its
environs as a matter of choice. The list of
potential positive reinforcements is long and has
too often been overlooked by city designers. Our
(Kikutake, Craven) list is as follows:

- Opportunities to derive adequate economic
  income

- Opportunities for social affiliation in
  recreation, organization, sports, spectator
  sports, conversation, theatre, arts, music,
  etc.

- Opportunities for achievement in organizations, in business, intellectual, and
  physical activities, in culture, the arts,
  etc.

- Opportunities to engage in dominance and/or
  submission with family or selected associates, or with pets or inanimate hobby
  objects

- Opportunities to engage in dogma and
  independent belief with family or selected
  associates, or with pets and hobbies

- Opportunities for privacy, individually, in
  pairs, or in small groups

- Opportunities for afferent reinforcement
  through experiencing spacious and unscarred
  natural landscapes and environments

- Opportunities for experiencing variety and the freedom to select or alter one's local environment

- Preservation of similar opportunity for posterity

Given this large number of conditions, one may well ask whether there exists any reasonable configuration which will permit all of the reward stimuli to flourish and at the same time protect against all of the aversive stimuli and meet minimum requirements for the mechanics of living. Quite obviously many of the reward goals are denied by modern urban-suburban living.

For example, the low density in the suburbs has made social affiliation on a non-planned, casual basis extremely difficult. In the city proper, crime in the streets is a real and unsolved problem, not only for the victims of this aversive stimulus but as an expression of unfulfilled needs on the part of the individuals engaged in this anti-social behavior. The fear of crime makes prisoners out of many apartment-dwellers, particularly the aged and the female, destroying their own opportunities for socialization. The need for companionship and domination is often expressed in the ownership of pets which, in turn, creates aesthetic and public health problems in the city streets. The security-conscious nature of the apartment complex, which denies door-to-door solicitation and is restrictive of unsolicited mail communication, isolates the individual from dogma-resolving organizations such as the church and various forms of the political process. Many other examples can be cited, but the reader who has any experience within the city can cite countless situations and times when these idealized reward structures are unavailable or are defeated.

Perhaps these reward goals are unachievable in the city and thus unrealistic or inappropriate for design purposes. If we believe that the city is organic (i.e., that its formation and growth processes are basically unplanned), then there will have been an attempt or attempts to adapt to an optimum form of social existence. If we can find (and we can) examples of stable, satisfied communities whose existence has not been transient, then

we have an empirical test of these design principles. Certainly, the evolutionary process which
produced these communities would have satisfied
the suggested principles of city design if the
principles are indeed appropriate and correct.

If we seek a solution at the smallest scale of
society, we can find it in man's Neolithic lake
societies. As proof of the fundamental and organic
nature of these societies, we note that Neolithic
communities of nearly identical form and size grew
and flourished around the world in temperate and
semi-tropical zones. We can understand the pattern
of development as man the hunter followed the small
game to the lake shore. There the seeds of the
fruits consumed by the herbivores were deposited
in their feces on the fertile muds of the edge of
the lake to initiate an annual cycle of seed time
and harvest. Nature thus provided the first agrarian hold on roving man, and communities developed
on the shoreline. But shoreline communities did
not meet the needs of society, (at least not as
outlined in this paper) and in order to survive,
the community had to move to the center of the lake.
The great historian Herodotus describes one of
these communities as follows:

Their manner of living is the following.
Platforms supported upon tall piles stand in
the middle of the lake, which are approached
from the land by a single narrow bridge. At
the first, the piles which bear up the platforms were fixed in their places by the whole
body of the citizens, but since that time the
custom which has prevailed about fixing them
is this: they are brought from a hill called
Orbelus, and every man drives in three for
each wife that he marries. Now the men have
all many wives apiece, and this is the way in
which they live. Each has his own hut, where-
in he dwells, upon one of the platforms and
each has also a trap door giving access to the
lake beneath; and their wont is to tie their
baby children by the foot with a string, to
save them from rolling into the water. They
feed their horses (sic)* and their other
beasts upon fish, which abound in the lake to

_____

*Probably "pigs", rather than "horses".

such a degree, that a man has only to open his
trap door and to let down a basket by a rope
into the water, and then to wait a very short
time, when up he draws it quite full of them.
The fish are of two kinds, which they call the
paprax and the tilon.

The description would be equally valid for any
one of the Neolithic lake societies whether in
Britain, or Switzerland, or Etruria, or Africa, or
Peru, or Thailand, or the Philippines. Examination of these early communities reveals how closely
they meet the design principles of our utopian
city. The economic supply of food was obtained
from the game of the high hills and the abundant
fish supply of the lake; water for drinking came
from mountain streams continuously flowing from
the winter crop of snow; clothing from animal hides;
and shelter from the logs of tall timbers felled by
acts of nature. Land transport was also facilitated by topography, it being necessary only to
drag the large game downhill to the water's edge,
there to transport by coracle or raft to the pile
dwellings. Communication by drum or smoke signal
was adequate for the day, and disposal of waste was
simply accomplished by squatting over the platform
opening and dropping into the lake. These simple
cultures knew, as modern societies know, that
drinking the waters of the lake was tabu; but they
also knew what the environmental superstitions of
modern man have suppressed: that swimming in
waters enriched by biological waste presents no
hazard to the health. These same untreated wastes
were rapidly converted to fish more than fit for
human consumption. Thus, the Neolithic lake
community met the minimum requirements for survival
in a handsome manner.

Not all of the aversive stimuli of nature and
man were avoided, but these communities were
remarkably free from many that afflict modern man.
We have already seen how one or two simple tabus
freed them of most of the pollution problems which
so afflict our day. Less obvious is the protection
from fire, flood, and earthquake afforded by the
waters of the lake from their properties as an
extinguisher, as a flood control reservoir, and as
a decoupling medium for earth tremor. The configuration of the community was ideal for defense,
particularly against invaders having no knowledge

of the art of swimming or seamanship. One can see
the attacker now, launching clumsy and ineffective
log rafts which are quickly tipped by the defenders
who adroitly grapple with the invader in the waters
of the lake. How quickly the attacker's grip
becomes that of a drowning victim rather than that
of a warrior in battle. One can see the impossi-
bility of a blockade, since by stealth of night the
entire periphery of the lake is available as a
location for landing and embarkation. Thus, defense
against the invader was simple. Crime and civil
disorder were also of minor social import as the
result of the rigid tribal and family structures of
these simple communities which visited punishment
on the iniquitous rapidly, simply, and rough-hand-
edly.

What really made these early-day Walden ponds
a stable and happy culture was the existence of
many opportunities for reward. The organization of
the platforms provided an ideal configuration for
social affiliation. Dwellings were in general
located at the periphery, and community and family
gatherings were in the center of the platform
configuration. Thus, the assembly process from the
low density family unit to the high-density community meeting was quickly and easily accomplished.
At the same time, dwellers on the periphery could
look outward and lift their eyes unto the hills,
experiencing the wonders of spacious and unscarred
natural landscapes, the snows of winter, the
greenery of summer, the foliage of fall. Opportunities for achievement in the hunt, or in sports
in clearings on the shore, or within the tribal
structure were abundant, and the community was
bound spiritually by religious dogma and symbolism.

When these communities failed, it was due to
their inability to preserve the opportunities of
the societies for their posterity. As Herodotus
pointed out, each man had many wives. Thus, each
man had many children, and with the growth of
population came the increase of platforms until
the lake was covered, and the vital functions of
defense, transportation, fishing, aesthetic reinforcement, etc., were severely compromised. The
solution for many of these societies was exile over
the rim of the crater which had given birth to the
lake, onto the streams that flowed inexorably down
to the sea. Thus, pieces of civilization were

forced from their stable and tranquil land-sea configuration to start life anew in the coastal zone.

Before we leave these tranquil societies (many of which survive to the present day) we note that these communities can be represented symbolically by a three-dimensional configuration having the following basic characteristics: the land is reserved for agricultural, recreational, and low population density functions of the society; the surface of the streams and the large mass of water is reserved for transportation. High density living takes place in three-dimensional beehive-type structures located above the water surface with lowest density on the outer periphery and highest density in the center. The technological sophistication of the society is to be found in these structures. The waters below the surface are utilized for industrial functions, for cooling, for processing of wastes, for manufacture of protein. The total configuration provides simultaneously the natural configuration for defense and a configuration which, blending with the environment, requires little in the way of maintenance and upkeep.

We may now ask whether modern technology can replicate the Neolithic model at a larger scale and, if so, what may be the current economically feasible limit. We know that our society enjoys the aggregation of 100,000 or more people in a single stadium for the reinforcement that mass participation in spectator sports provides. Thus, concentrations as high as 5,000 people per acre are desired. On the other hand, for open space enjoyment, concentrations of less than one person per acre are desired.

These density dilemmas can be resolved by the physical organization of society in a three-dimensional community which is purely pedestrian or which at most employs simple people-movers (escalators, moving walks, golf carts, etc.). If, for example, we desire to collect an audience of, say, 10,000 people for a concert from a population of, say, 100,000 who are distributed in a circumferential belt having the conference hall as a center, then one must travel some four miles before the suburban density of 10 to 20 people per acre is obtained. On the other hand, if a truly three-

dimensional city having a height of 20 stories were
to be constructed, then this density would be
achieved at the healthy walking distance of one
mile.

To understand this concept more fully, we
should note those activities which require a view,
those which do not, those which admit of high
density, and those which do not. In the first category are theaters, concert halls, basketball
arenas, even baseball and football stadiums,
churches, classrooms, political rally arenas, etc.
Of lesser density, but nevertheless high, are
department stores, grocery stores, cleaners, service
establishments, restaurants, hospitals, medical
centers, business offices, barber shops, beauty
shops, etc. Lower density goes with living spaces,
and lowest density with parks and airports.

Conceptually, this leads to a beehive concept
in which living spaces are on the outer skin, non-
visible one to the other, and each with a view of
an unencumbered environment over a solid angle por-
jected from the viewer and limited only by the
collimation required for privacy. The individual
travels internally into the hive toward successively denser areas of population: at the outermost
level for socialization in plazas and courtyards;
at an inner level for shopping and services; at a
further inner level for offices and business and
vocation; and at the core for entertainment and
convocation.

To meet low-density needs of the population,
the resident of the complex travels externally from
the beehive to parks and open spaces to a transportation system which can carry him to the distant parts of the environmental park which surrounds his high-density complex. Ideally, this
transportation system should be placed below the
"beehive" so that it is equally accessible to all
citizens; or (as is done at Disneyland) a peripheral transportation system could be provided to
carry individuals to the desired outbound radial.

The model as thus far described leaves out the
space and locational requirements for industry,
waste disposal, communication, power, warehousing,
movement of freight. If these could be located
below the beehive, then an arrangement idealization

is obtained which in theory would allow the full design goals of the city to be met. This concept has been explored by Paolo Soleri in his arcology concept.

The realization of the Soleri concept presupposes that the arcologies which he has designed are structurally feasible within the current state of the art and that a transportation system is provided which can move people from the high density core of the city to the low density parkland periphery. The realization further presupposes that the energy and capital costs will be low enough so that the community is economically viable in competition with other forms of social organization.

If we first examine the transportation, we note that technology now allows three major alternatives: the motor car, the fixed guideway rail system, and waterborne riverine and canal transport. Despite its apparent flexibility, we note that the automobile fails to permit the high density aggregation which is desired for spectator events and it equally denies the opportunities for the low population densities required for wilderness enjoyment. This paradox results from the fundamental character of automotive transport. If the automobile is employed (as it is) for transport to high population density complexes such as athletic stadiums, the attempt to achieve high density is defeated by the extensive fields of parking lots and networks of highway access systems which are required to accommodate the population to be assembled. This limits population density to not more than 500 people per acre. It further requires a separation between shopping centers, theatres, and stadiums so that each may accommodate its own burden of automobiles during the period of maximum utilization. As a consequence, the time pollution resulting from transportation to each complex limits the number of activities which may be enjoyed in any given interval of time.

At the other end of the scale, access to each wilderness area requires the installation of highways, power lines, communication lines, etc. -- a total umbilical package whose economic cost cannot be justified by the low density use. Soon the installed umbilical is loaded with parasitic

attachments along its entire length until the maximum economic utility of the umbilical is realized.
At this point, the desired low density of population in the wilderness is lost (we might call this
the Yellowstone Park phenomenon).

The replacement of the automobile by fixed
guideway mass transport systems will be effective
in obtaining higher population densities. The
effectiveness of the New York City subway system in
this regard is undoubted. For example, with the
combined resources of the BMT, IRT, 6th Avenue, and
8th Avenue subway systems, it is perfectly feasible
to move 100,000 or more passengers per hour into
and out of the Times Square area of New York. This
would be accomplished without obstruction of the
street level surface by any vehicle. On the other
hand, the higher lineal cost of fixed guideways
makes the use of this form of transport into wilderness areas one of extremely high cost. Nonetheless
we must conclude that the use of fixed guideway
subway transit systems will satisfy the arcology
concept at a scale beyond that now realized in the
highest density cities.

If we examine the third alternative of utilization of existing water courses and canals supplemented with a modicum of canal construction, we are
surprised to find that they are plentiful, occur in
the very center of the areas of highest population,
and are connected by means of bays, estuarine, or
coastline transit to land masses having little or
no population density. This "free" and interconnected fixed guideway system has the further
advantage that vehicles of high passenger capacity
are easily borne on the water surface. As a consequence, the limitations of the automotive and the
fixed guideway system are not present. If, for
example, it is desired that transport to wilderness
areas be effectuated, the total capital investment
is simply that for the most rudimentary of landings.

In the high density portions of our present
city complexes, the existence of waterways is
merely a concomitant of the drainage which is
required for the protection of these high density
areas.

To understand how convenient and effective
these natural corridors can be, the population

density which is available to them in the American city should be quantified. The author has analyzed a number of cities in terms of their ability to benefit from marine mass transit. Population and other demographic data were taken from the Atlas of Twenty American Cities. The usable waterways were identified, and one-quarter and one-half mile swaths were drawn on either side of the waterway. The population within these swaths was then identified and calculated. Substantial percentages of the total population were found to live within these corridors, as follows:

Table 1

	Percentage of Population Living on Either Side of Usable Waterways	
	1/4 Mile Swath	1/2 Mile Swath
New Orleans	27	49
Seattle	25	45
New York	19	39
Honolulu	22	39
Boston	18	34
Philadelphia	12	28
Chicago	9	21
San Francisco	9	18
Cleveland	4	12
Washington, DC	4	12

The population's access to these waterways compares favorably with--and is in many cases superior to--the access to carefully designed subway systems. Of particular significance is that the population density per lineal mile is high and, in most cases, quite uniform along the waterway. A comparison of New York and Boston systems shows a population density of 79,000 per mile* for New York subway systems; 25,000 per mile for New York City waterways; 19,500 per mile for Boston waterways; and 13,600 per mile for Boston subways. (Figure 1 shows accessibility to New York waterways.)

*This figure (79,000) is too high in terms of "population per mile of subway systems" because for about 70 percent of the distance, more than one subway cuts through the swath.

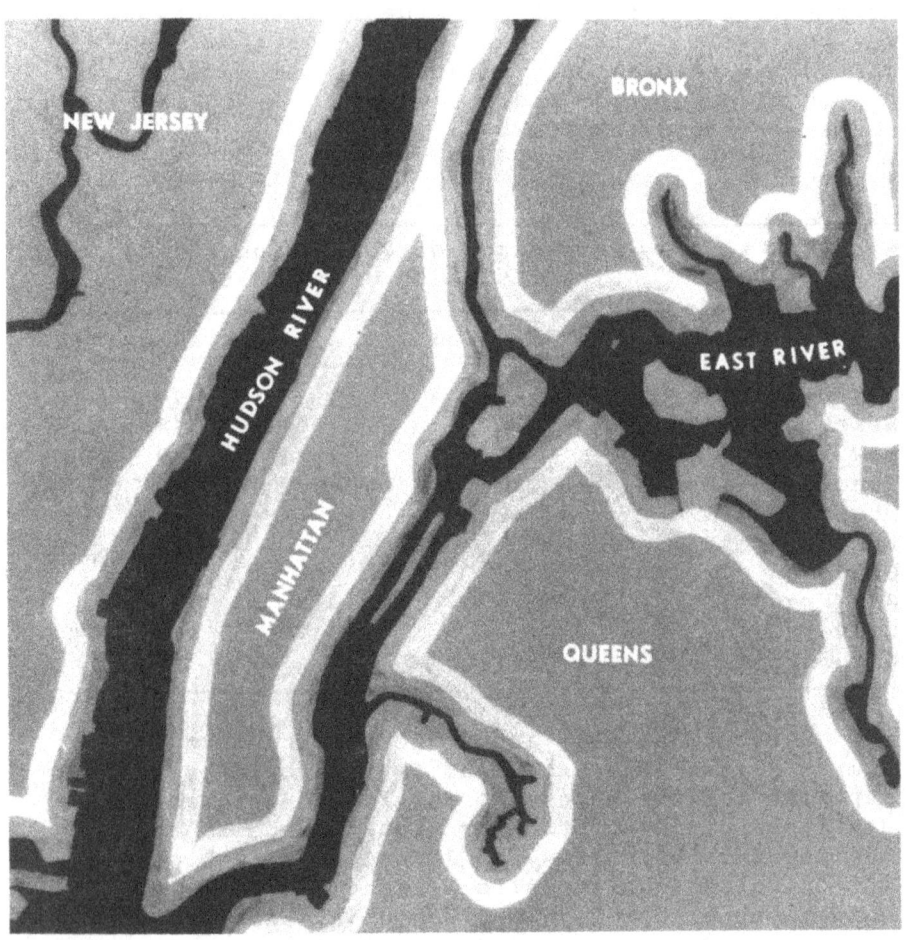

Figure 1.   Accessibility of New York to waterways
usable for marine mass transit (¼- and ½-mile swaths).

The evidence is quite conclusive that existing waterways have the potential to serve as conduits for public transportation in vital transportation corridors. The utilization of the waterways for revitalization of the _entrepot_ functions of the city appears attractive for many cities. Integration of container barges into the highway networks is relatively easy because of the ubiquitous nature of the net. Revival of the barge for railroad car transport is feasible (should there be any railroads financially able to construct them). Even a cursory examination of subway and fixed guideway systems indicates that nearly all have stations on or near the waterfront in locations that would make interchange construction relatively economical. Interchange with airports is relatively easy in many cities. As previously indicated, and for a multiplicity of reasons which include noise pollution abatement, glide path considerations, availability of filled land, etc., many city airports are located at the edge of the ocean or alongside a major waterway. Thus airports in New York, Philadelphia, Washington, San Francisco, Cleveland, Los Angeles, Boston, Honolulu, have immediate access to the water network.

Given the convenient location of most waterways in the modern city, the next question is, of course, the availability of modern water-based vehicles to act as the couriers on these systems. A fairly substantial number of such vehicles have been developed. Unfortunately, many are unavailable in the United States because of the prohibitions of the Jones Act and the absence of an adequate small-craft shipbuilding capability. It is for this reason that many in the United States are unaware of the revolution in small-craft design and development that has taken place in the past two decades.

Given conveniently located waterways, the next question concerns the availability of modern water-based vehicles which can carry people in quantities to truly qualify as "mass transit," and which have a ride quality in the heaviest of seas in which they will operate such that motion sickness does not result. There are such vehicles. Unfortunately, the regulatory and socio-economic processes of the United States have exiled all of these craft from our shores. The major break-

Figure 2. Mitsui Engineering & Shipbuilding Co., Ltd. Semi-Submersible Passenger Ferry.

through in marine ferries is the semi-submersible
platform ferry. This craft, initially developed
and currently ignored by the United States Navy,
is being introduced into commercial passenger ferry
service by Mitsui Engineering and Shipbuilding
Co., Ltd. (Figure 2).

The craft consists of twin submarine hulls
whose draft is adjustable by ballasting. Twin
struts pierce the surface and support a box-like
platform above the free surface. The craft has a
number of significant characteristics. The chief
one, of course, is stability at any speed in a seaway such that motion sickness will rarely occur,
and on those rare occasions, the sickness will
probably be more psychological than physical.
Because the hulls are well below the free surface,
the wave drag on the craft is reduced, and a significant speed advantage for a given power can be
obtained. Another major advantage is the ease with
which it can change draft to accommodate shallow
water or to mate with a pier. As such, the craft
can play a major role as an interface between canal
and river systems and systems which must operate in
open and exposed waters. A major advantage may be
that of all river craft, the semi-submersible is
perhaps the most suitable for roll-on roll-off
movement of automobiles. Because of the shape of
the hull and because of the inability to substantially change draft, all of the other vehicles are
inappropriate for this function. The hull shape
makes location and stowage of the vehicles for
transit a cumbersome and slow process, and the
inability to change draft requires a complex and
cumbersome loading ramp facility. The upper structure of the semi-submersible can be arbitrary in
shape and is in general box-like or shaped like a
parking lot or a multi-tiered parking lot. Regardless of the load, adjustment of ship's draft
through movement of ballast will permit the semi-
submersible to mate with the access road.

The semi-submersible can be built in any size.
Thus, 2,000-passenger craft having displacements of
less than one ton per passenger are feasible for
conventional ferry operation anywhere on the
world's oceans. At modest speeds, the energy cost
per passenger per mile will be among the lowest of
transportation costs for any mode of transport.
Low in cost of production, low in use of energy,

high in passenger load capability, and competitive
with land mass transit in speed, the semi-submersible represents a fulfillment of the economies of
scale and an opportunity to expand by an order of
magnitude the passenger-moving capabilities of mass
transit systems.

A complement to the semi-submersible is the
high-speed hydrofoil. The most sophisticated embodiment of this form of craft is the Boeing hydrofoil.
This craft literally flies in the water on underwater foils. The angle of attack of the foils is
adjusted to compensate for the waves and a comfortable ride is attained in all but the highest sea
states. The 300-ton, 200-300 passenger craft was
technically successful but financially unsuccessful
in inter-island service in Hawaii. Overseas, they
are financially and technically successful, particularly on the run between Hong Kong and Macao. The
dollar deterioration has made them attractive for
use in Japan.

On sheltered waters many alternatives are
available. The long-time standard of comfort and
convenience and volume transport remains the New
York City ferry which carries 2,000 passengers and
many vehicles. More modern versions ply the waters
of Puget Sound. In warmer waters, the Vaporettis
of Venice and the riverine hydrofoil systems of the
Soviet Union have demonstrated the capability of
high-speed, high-volume, comfortable, and low-cost
mass transportation from the very heart of the
dense metropolis to the most remote unpopulated
areas without additional capital investment in
roads or fixed guideways.

The outlines of a sea-based solution are,
however, inherent in current developments in the
industrial uses of the sea. An evolutionary process which is progressively producing a seaward
extension of industrial and urban systems is
rapidly unfolding. The thread of progression of
significance in developing a sea-based city complex
is as follows: 1. the conceptualization, design,
and construction of the dynamically-positioned
semi-submersible platform, 2. the conceptualization
and concept design of the pre-stressed concrete
floating city module, 3. the conceptualization,
design, and initial stages of construction of the
barge-mounted offshore nuclear power plant, 4. the

conceptualization, design, and construction of the
pre-stressed concrete EKOFISK oil storage facility,
5. the symbolic demonstration of the floating city
concept at the 1975 International Ocean Exposition
at Okinawa, utilizing a semi-submersible as the
demonstration platform for the AQUAPOLIS concept
(Figure 3), 6. the conceptualization, design, and
construction of the "verticle bottle" semi-submersible as employed in the CONDEEP oil storage
tanks, 7. the conceptualization, design, and
construction of the barge-mounted floating industrial plant, 8. the conceptualization and concept
design of the barge-mounted and semi-submersible-
mounted floating coal power plant (Figure 4), 9.
the conceptualization and initiation of development
of the ocean thermal energy conversion plant.

This set of developments makes clear that the
economically feasible urban platform is within the
current state-of-the-art.

These specific milestones are just a few of
the total set of offshore platform developments
which have been initially associated with the oil
industry and are now increasingly associated with
oil, energy, and industry. The offshore development was initiated by the development of the offshore fixed platform whose success was followed by
a bewildering variety of drill ships, jack-up
barges, and semi-submersibles. In deep water,
fixed platforms or moored systems became progressively non-feasible. It was a discovery of the
National Science Foundation Deep-Sea Drilling
Project that dynamic positioning (the use of
rotatable propellor systems to counteract drift)
was feasible and required relatively little energy
in most oceanic situations. This resulted in the
design of a structure called the MOHOLE platform.
Although never completed, it became the prototype
for a large number of semi-submersible oil rigs.
Such platforms like the semi-submerged ship consist of twin submarine hulls connected by vertical
columns to a box-like structure located well above
the free surface. Insofar as distribution of
usable space is concerned, it matches that desired
for the idealized three-dimensional city module.
Although these platforms are far more stable than
conventional ships, their commercial function has
not required the freedom from sensible motion that
would be required for urban uses. Similarly, the

Figure 3. AQUAPOLIS, the demonstration floating city at the 1975 International Ocean Exposition, Okinawa, Japan.

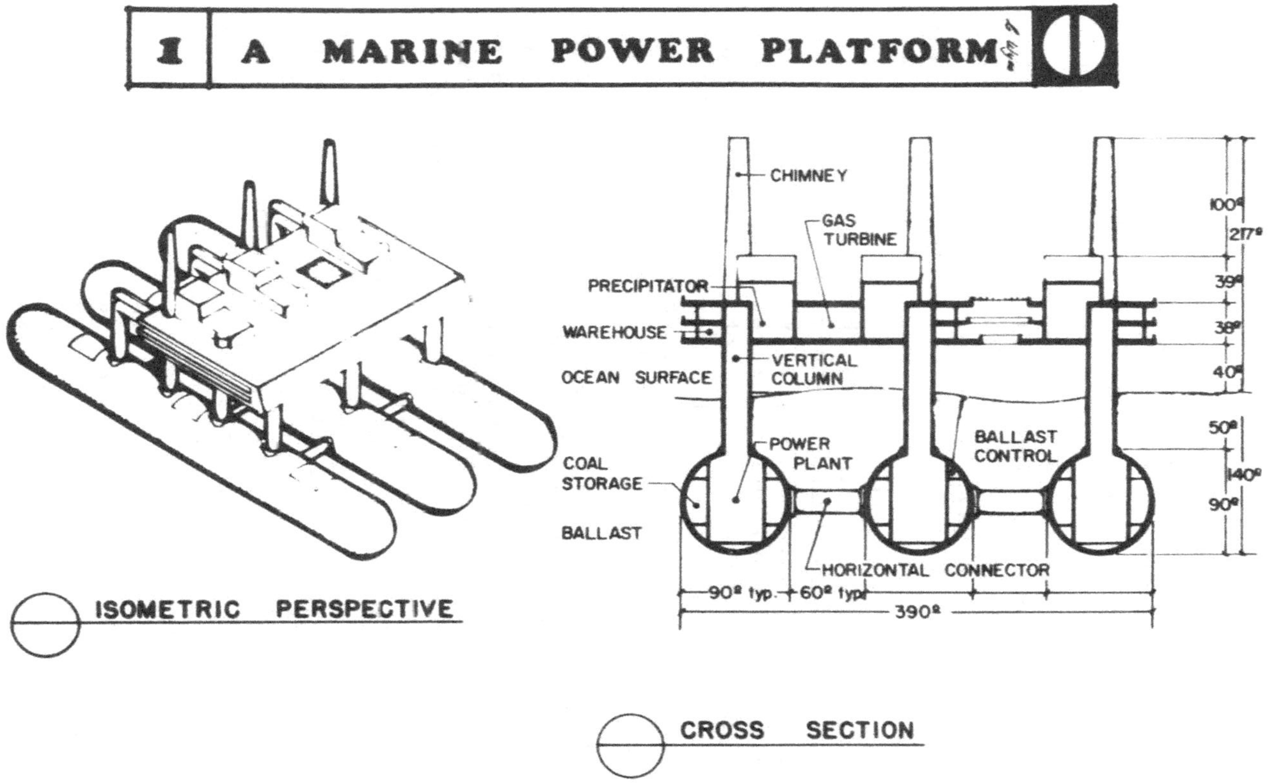

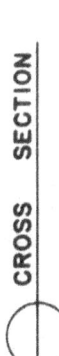

Figure 4. University of Hawaii floating coal-fired power plant.

volumes of usable space which are far greater than
required for their commercial function are nonetheless required for urban purposes.

It was, however, the recognition of these
characteristics that led to the basic design of the
Hawaii floating city concept (Kikutake, Craven,
et al.). In this conceptualization, the stable
platform configuration consists of a triad of vertical column structures which are fully submerged.
Each of these columns has a column of very much
smaller diameter superimposed on top which penetrates the free surface, and supports a multi-deck
structure located well above the free surface.
Particular care was taken to find the theoretically
most efficient structure for maximum stability in a
seaway (Seidl) and, as a result, the required
"motionless" configuration is obtained for platforms of large but nonetheless reasonable scale.
At the time of this design, it was postulated that
the structures would be of pre-stressed concrete
although this technology had not yet been proved.
It was also postulated that power plants and industrial systems for offshore platforms were economically and technologically feasible.

Contemporaneous with the floating city design
was the conceptualization and design of the
Westinghouse Corporation's floating nuclear power
plant. These barge-mounted plants are to be located along the Eastern seaboard in the relatively
shallow waters overlying the continental shelf.
The power generated by these plants is to be
carried by underwater cables to the user. Calculations indicated a large number of superiorities of
these plants over their land-based counterparts as
follows: 1. reduction in costs due to multiple
production in shipyards, 2. reduction in costs and
increase in safety due to elimination of structural
foundations, 3. reduction in costs and increase in
safety by elimination of hazard of earthquake,
4. increase in safety and efficiency of the city by
the location of the power plant at a considerable
distance from the user, 5. increased safety from
sabotage by virtue of remoteness, 6. aesthetic and
non-polluting by virtue of remoteness, 7. low in
cost by virtue of sea-based supply, 8. protected
against storm by a large dike, the plant is virtually immune to natural catastrophe.

On the basis of their study, Westinghouse Corporation has proceeded with the construction of a "shipyard" for the sole purpose of building the floating plants and with the contractual and permit process to build the first of these units. Unfortunately, the regulatory, socio-economic, and self-punitive characteristics of our current social and political organization have virtually prevented the completion of this project. It nonetheless demonstrated that floating nuclear power plants (whose technical feasibility was demonstrated in the nuclear submarine) are commercially feasible.

The conceptualization of the floating city was also contemporaneous with the conceptualization, design, and construction of the EKOFISK offshore oil storage facility. This major innovative development which revolutionized the use of concrete for offshore structures was constructed in Stavanger, Norway. The tank, designed to store one million barrels of oil, consists of nine nested cylinders approximately 80 feet in diameter and 280 feet tall (the height of a 20-story building). The cylinders are made of pre-stressed concrete. The entire assembly is surrounded by a concrete wall which is perforated in a manner that will most effectively absorb the energy of storm waves. The facility was built by the initial establishment of a dewatered cofferdam. The base was poured in this area and the vertical columns constructed by a slip form technique until they were tall enough to be floated. The cofferdam was then removed and the installation floated into deeper water. Construction was completed by continuation of the slip form technique while the structure was floating in deep water. It was then towed to the site and sunk in place. Although visited by a major fire, the EKOFISK facility has successfully demonstrated the ability to build large volume structures in sheltered bays and to tow them for use in the open ocean.

The Hawaii design pointed the way to a symbolic demonstration of the floating city concept of the 1975 International Ocean Exposition at Okinawa. Although contemplated, time precluded the construction of a dynamically positioned vertical column platform. A more conventional semi-submersible moored rig was chosen to represent the floating city module. Significant features included the

dual location of the AQUAPOLIS either connected to
the land by a fixed bridge or located at sea and
accessible only to marine transportation. It could
also be repositioned well at sea should a typhoon
of extreme intensity be anticipated. A representative set of city services were provided, which
included power generation (fossil fuel), waste
treatment, fresh water production, etc. Another
significant demonstration was the fact that the
AQUAPOLIS was constructed at a shipyard in the main
islands and located in Okinawa with a minimum of
site preparation. The AQUAPOLIS did not have the
requisite stability (i.e., visitors could sense
slight motion of the platform), the volume of
usable space, or the low cost per unit volume that
would be required for an economic urban module.

Fortunately and once again contemporaneously
with the development of AQUAPOLIS, was the conceptualization, design, and construction of the
CONDEEP oil storage facilities. These pre-stressed
concrete structures are of the size and configuration which were postulated as ideal for the floating city module. Although they are not dynamically
positioned in their location as storage tanks, they
were nonetheless built at a site remote from their
final location and were effectively towed from the
construction area to the site location. These
situations demonstrate beyond the shadow of a doubt
that massive pre-stressed concrete elements can be
built in a size and configuration that can accomo-
date the volume and arrangements requirements of
the ideal three-dimensional city complex.

Although the CONDEEPS go far in demonstrating
the feasibility of structure, they do not demonstrate the feasibility of location of industrial
facilities at sea. Two studies were conducted in
Hawaii on the feasibility of location of conventional coal power plants. One study postulated a
barge-mounted plant, the other a semi-submersible
platform. Both indicated that the cost of such
a plant would be the same or less than the cost of
a land-based plant of similar capacity. The advantages of sea-basing which were discovered in the
Westinghouse nuclear plant design were found to be
equally applicable to the conventional power plant.
In addition, and depending on plant siting with
respect to prevailing winds, it was found that the

polluting aspects of coal-fired plants could be
substantially minimized.

These studies have been effectively confirmed
by the actual construction and installation of
barge-mounted industrial plants which have been
built in Japan and installed in far distant overseas locations.   Ishikawajima-Harima Heavy
Industries has completed plans for barge-mounted
steel mills, cement plants, thermal power plants,
a utility plant, a refinery, a sugar plant, and
a desalinization plant and has completed and installed a pulp plant with its associated power
platform.   The plants were built in Japan and
installed in the Amazon.   The journey was antipodal
and thus demonstrated the global capabilities
inherent in this form of construction.

The sum of these developments is such that it
is now technically feasible to evolve or develop an
urban community which matches our ideal model.
This is illustrated by a configurational layout on
a typical coastal area on the west coast of Japan.
A small island offshore serves as the locale for
the local airport.   Industrial facilities, power
generation facilities, and waste disposal facilities are located on separate and separated floating complexes.   The urban and industrial center is
located on a slowly mobile platform at the mouth of
the valley.   All of these are connected by marine
mass transit to the interior along the streams of
the valley.   Farms, agriculture, schools, stadiums,
government plazas are located in the lower valley,
and in the upper valley, the temples, parks, and
teahouses of an earlier era are preserved and made
available.   Note that little, if any, structures
are located on the coastline and that the coastal
zone itself is relatively uncluttered except for
recreational uses.   The concept fully recognizes
that the coastline is the delicate and fragile eco-
system and that the land above the highest tide
level is moderately robust and that the ocean
beyond the bay and estuary is very robust.   In
general, the choice for location of each element of
the urban system is between the land well within
the valley or delta or the sea a mile or more from
the coast with both complexes tied together by a
transportation system which is orthogonal to the
coastline and which penetrates to the uppermost
reaches of the navigable streams.

If a case has been made that it is possible to
build a large-scale urban complex that is economic
and functionally beautiful in the sense that it pre-
serves small-scale phenomena and satisfies the
design criteria for the psychologically satisfying city, one may also ask the more trivial question as to whether such complexes can be aesthetic
from an architectural sense.

Except for the AQUAPOLIS, the author is unaware that an architect in the classical sense has
ever been employed in the design of any of the offshore structures now extant. (The preliminary
refinery plant on an island in the harbor of Long
Beach, California, is another exception. The plant
camouflaged to look like highrises on a palm tree
island can be deemed aesthetic or not, depending on
your attitude toward highrises.) Two factors have
obviated the need for such architectural treatment.
The first is the fact that the major portion of the
offshore structures are submerged and therefore not
visible. Observers directly above can see with
clarity no further than 50 feet in the clearest of
water and even then the view is obscured and enhanced by the filter of blue. The second is the
fact that platforms at a distance from shore soon
fall below the horizon and are not seen except for
those few who man the facility or visit for purposes of inspection.

An urban complex will, of necessity, be close
enough to shore and will have enough of its structure above water to warrant architectural treatment. Such considerations were paramount to the
Hawaii floating city design. A prime factor in
aesthetic enjoyment was the mobility requirement
which would have resulted in the complex moving
back and forth (albeit very slowly) in a more or
less prescribed track, and rotating. Thus, the
view from shore and the view from sea would be
constantly changing. A dweller in a peripheral
apartment would have an ephemeral view--on some
days, the view would be of sunrise, on others of
sunset; on some days of land and on others of sea.

It should not be difficult to make the overall
profile as seen from a distance quite attractive.
Many artists have found the seascape to be a subject of great appeal and have found it convenient
to employ floating objects as the centerpiece of

the masterwork. In general, these views have been
pyramidic with the water serving as the base of the
pyramid, the structure serving as the lower base,
the superstructure as the upper portion, and the
sun or moon as the single illuminator located at
the apex.

The Hawaii floating city configuration was
designed with this concept in mind. Although it is
true that most architectural models and drawings
are beautiful whether or not the prototype proves
to be so, the floating city model leaves little
doubt of grace and beauty (Figure 5). The webbed
curtains which are employed for screening sun, and
as a windbreak, provide the graceful flowing surfaces which carry the eye from the horizontal
surface of the water to the delicate minarets. No
structure above the surface is more than eight
stories tall, nor need it be, for the bulk of the
usable volume is located below the surface. The
total design, though big, is beautiful; it is
functionally complete, a concept of land and sea,
a thing of beauty, and a joy forever. It is to be
hoped that someday the regulatory, socio-economic
and self-gratification structure of the United
States will permit the aggregation of capital and
the unfettering of imaginative and competent
architects and engineers so that this concept and
others of its ilk can be developed for the better-
ment of man.

Figure 5.   University of Hawaii floating city model.

Overview

*American agriculture is on the wrong track.
The farming methods of "agribusiness", the policies
and programs of the federal government and the
states, and the exaggerated use of chemical fertilizers are leading to massive erosion of topsoil
and the destruction of the soil's biological
resilience. The emphasis should shift to ecologi-
cally-sound farming methods and to the restoration
of the family farm. James A. McHale argues that
changes in the institutional policies and the technical methods of American agriculture are necessary
to insure the nation's ability to produce abundant
and healthful food in the future.*

James A. McHale

11. The Future of Rural Technology

Superficially, the American agricultural system appears to be one of our great success stories. In an age when 500 million people go to bed hungry each night, the United States is able to feed its own populace (perhaps too well) and export 25 billion dollars worth of farm produce annually. Farming is just over 3% of the gross national product, but last year it produced 20% of all our exports. But that production was achieved through our past and present agricultural land and food policies which encouraged agribusiness at the expense of the farm family, the environment, consumers, and our soil. This is, I believe, one example of "big but not beautiful."

My thesis is that we desperately need a program that will recapitalize and revitalize Rural America and which will encourage ecologically sound farming methods, preserve threatened agricultural land, stimulate rural and small town development, and insure sound nutrition in foodstuffs for man and animal.

I want to tell you, "Things are not rosy down on the farm." There are fewer than 3 million farms today in the United States, half the number in 1950.[1] Farm debt is $136 billion and debt service takes 10% of all farm gross income. Farm income is at 70% of parity.[2] We have sent 30 million people to the cities in the last thirty years, and many have not found a better life there. Many of these displaced rural families have contributed to the increase in welfare payments, unemployment, and inner-city destruction. We still are losing rural population at a rate of 700,000 a year. If we continue to lose farm families at the present rate, we may well be <u>minus</u> the small farmer long before the year 2000.

We need the family farmer and the production philosophy he represents to maintain a quality of life for Rural America and to produce a better quality of food for all Americans. The economic pressures which squeeze the family farmers out result in corporations becoming increasingly involved in every phase of our food production and distribution system. (Fifty corporations now control 75% of all food manufacturing assets.)

Agribusiness as presently structured has not been an unqualified success for the rest of the U.S. population. Consumers face steadily increasing food prices and declines in the nutritional value of food. It has been estimated that the American diet contains 166 pounds of sugar and 9 pounds of food additives annually. Add this to the 5 pounds of herbicides, pesticides and fungicides that are manufactured each year for each man, woman and child in the United States, at least some of which are consumed in the final product, and I think the case is very strong for changing our direction in agriculture at once.[3] Indeed, there is some speculation among scientists and doctors that some of these chemicals, while not affecting our health in ways that we immediately notice, might cause genetic change and damage in future generations. There are other problems also: we all know the DDT -- 2,4-D -- 2,4,5-T -- PPB stories, but we don't know the final results. Let's hope Rachel Carson's "Silent Spring" doesn't come true.

I feel that we are racing against time. I can't visualize the United States remaining a viable nation unless we preserve and encourage relatively small and independent business entities, including the family farm. To do this we need a rational land policy, and the first step is to find out, "Who owns America -- Pennsylvania, Maine, Texas, California, and the rest?" In fact, twelve land owners own 52% of the land area in Maine, 25 own half of California; and we even see the United States government encouraging foreign investors to buy American farm land.[4]

The loss of our millions of farm families was

not by accident; rather, our cheap food and agricultural policies systematically favor the large corporate farms through tax breaks, credit provisions, and research geared to the "bigger is better" philosophy without regard to long-term impact on our environment, the farm family, or our rural culture.[5]

All this has been supported and nurtured by our U.S. Department of Agriculture, land-grant colleges, Extension Service, and some state departments of agriculture, in the name of progress, free enterprise, and supply and demand. The direct and indirect influence from agribusiness has been substantial if not controlling.

The Hatch Act of 1887 instructed our land-grant institutions to promote "a sound and prosperous agriculture and rural life," a permanent agriculture, and it says their purpose should be the development and improvement of the rural home and rural life. As far as I am concerned, the land-grant college experts have presided over the destruction of what they were mandated to preserve. They function in what has become a closed, self-serving society. I believe that the land-grant institutions have used the wrong standards. They have used industrial standards concerned only with speed and efficiency and have applied them as standards for agriculture. These are short-term concerns and even in industry they are not the only appropriate concerns. In agriculture where we are working with renewable resources such as soil that must be maintained and conserved, such standards are clearly inadequate.[6]

We should have learned from history. We have seen nations fall when they lost their renewable resources of topsoil and forest land. We have seen the Sahara Desert moving south at the rate of several miles a year after the forests were destroyed and the eco-balance lost.

We in the United States have been able to destroy a lot of our land in fifty years. There are those among us who remember the dust bowl days of the Thirties, when Oklahoma blew over New York City, and we are still losing our top soil. Through water erosion alone, our top soil is being drained at a staggering rate: nationally, erosion

averages nine tons per year, more than twice as
fast as the soil can possibly be replaced. This is
occurring not only on our hillsides, but also on
relatively flat, rich farmland where the American
farmers are following intensive planting methods
without the requisite intensive soil preservation
practices.

Let me quote the following excerpts from a
Government Accounting Office report to the
Congress -- CED 77-30, February, 1977:

> The Council for Agriculture Science and
> Technology reported in January, 1975, that
> more than 1/3 of the United States cropland
> was suffering annual soil losses in excess
> of the limit at which soil productivity
> can be sustained over time. The report
> suggests that we are less effective in con-
> trolling erosion today on some lands with
> severe erosion problems than we were 15
> years ago.[7]

An April 1972 Iowa State University research
report[7] stated that the United States was losing
4 billion tons of top soil a year through water
erosion, compared to 3 billion tons in 1934. The
report said that it would take a train of freight
cars about 633,000 miles long to move 4 billion
tons of top soil--a train long enough to circle the
earth 24 times! The report also said that today's
farmers were losing an average of 12 tons per acre
annually through water erosion, as compared to 8
tons in 1934. When I am talking about a ton of top
soil, I am talking about a layer the thickness of
one sheet of typing paper, spread over an area of
one acre. Another regional example of erosion is
that the equivalent of a 120-acre farm floats down
the mouth of the Mississippi River alone each day.

In June 1974, the Soil Conservation Service
reported that Iowa had experienced its worst spring
for soil erosion in 25 years. It stated that many
fields had soil losses of 40-50 tons an acre and
that some steep slopes lost as much as 200 tons
an acre.

As you can see, the extent of soil erosion in
America is massive, but the loss of soil nutrients
by erosion is even worse than the volume of erosion

would indicate. In New Jersey, for instance, soil materials removed from test plots by the selective action of erosion contained four to seven times as much organic matter, three times as much phosphorus and 1.4 times as much potassium as the original soil in the plots.[8]

The GAO report concluded that the rates of estimated soil losses revealed by their review, in combination with other reports, indicate that demise of the nation's agricultural productivity potential is possible.

And what has been done with the report? As far as I can see, it has been tucked away with the mountains of other GAO reports. It takes guts, imagination and work to make things happen, and we seem to act only in times of crisis. I am sure that when America wakes up to the fact that our lives depend on that few inches of top soil, we will respond. However, I am also sure President Carter and his budget-cutters will propose one more time to cut the Agricultural Conservation Program budget from the USDA budget. FDR had foresight and concern for the American farmer and our soil. His budget had $500 million for ACP--cost sharing--for farmers in the depression years of the Thirties, for conservation measures. Now ACP has been a mere $190 million of inflated dollars for several years running.

Clearly, we need a new approach to soil conservation. We have oil depletion allowances to encourage oil production. We also need a soil depletion allowance for our farmers to be used for long-term conservation practices to rebuild out land. I think a Des Moines reporter summed it up well a few months ago when he said, "Economics and politics have combined to block any real solutions to these environmental problems associated with American agriculture."[9]

I believe the evidence is there that federal and state agencies have mismanaged our soil conservation programs; Congressmen have invoked parochial concerns to block effective curbs on farm-caused pollution; and the farm organizations themselves have been complacent and even hostile to government attempts to deal with the situation.

It was Secretary of Agriculture Earl Butz who ended our land conservation policies in the early 1970's and urged our farmers to plant fence-row to fence-row, then to tear up their fence-rows and shelter belts and plant those areas too. This increased production helped the Russians in the famous Russian Wheat Deal. It also helped our balance of trade deficit. It improved farm income in the short-run—we sold the grain—but now our fragile, abused soils are washing down our slopes into our already polluted rivers.

Our present system of macro-agriculture is ruining our soils in another way. By encouraging the increasingly heavy use of commercial fertilizers, notably nitrogen, the soil biota vital to fertile soils are being destroyed.[10] Agriculture still depends heavily on nitrogen fixation, but our present agriculture practices using anhydrous ammonia fertilizer are destroying this possibility. Only a few spokesmen in our agriculture research and educational establishments are aware of or are concerned about the decline of the organic matter in our soils.[11]

Unfortunately, we have no reliable current information on the fertility of American agricultural lands. According to the GAO report cited above, the priorities for money use were misguided, and projects that increase production at the expense of conservation practices were funded. Also, their survey of the Soil Conservation Service functions pointed out that much of the technical information actually gathered is unwanted and unused by the farmers. What is needed is an outreach program to inform and educate the farmers on real, practical, workable soil conservation practices. And certainly, it is necessary to coordinate the efforts of the different USDA agencies, so that all are going in the same direction; instead of the county agents, Agricultural Stabilization and Conservation Service, and the Soil Conservation Service all coming up with different and conflicting recommendations.

Let me be more specific about what USDA has not done:

1. USDA has not collected data on the humic soil levels for some 30 years.

2. Most research on organic levels of our
    soils is over 10 years old.

3. No one has been working specifically
    on the question of numbers of soil
    microflora, and there is no reliable
    information on the national impact of
    chemicals and monoculture farming on
    beneficial soil biota levels.

4. Methodologies for determining soil
    tilth are non-existent.

I, for one, must ask, aren't the agencies concerned about the above questions? Or do they feel
that our soil is only for holding a seed and plant
in place, to be chemically injected, not naturally
nurtured?

Our agricultural research establishment has
successfully contributed to high production but has
ignored the soil, the basis of that production.
The future of soil productivity is threatened, and
we abound in ignorance concerning the second-order
effects of the miracle chemicals, particularly
anhydrous ammonia. We appear equally ignorant of
the causes for reduced nutritional value of our
food.[12] The current increase in pest infestation
and crop disease is also a serious problem. We
have raised hybrid bugs so strong that we can't
kill them between two flat boards. Everything is
out of balance.

An acre of fertile topsoil contains eleven
tons of biological life. Modern farms using toxic
technology often have less than two tons. Fertile
soils have as many as five million earthworms per
acre. Modern farms using acid fertilizer have
less than 100,000 worms per acre. The natural
ability of the soil and its organisms to decompose
organic matter has been substantially impaired.
Farmers are finding cornstalks that they plowed
under three years earlier still showing no signs
of decomposition. Our soils are suffering from
thirty years of toxic chemicals upsetting the
soil's eco-system and our total disregard of the
long-term effects of these chemicals on soil,
plant, animal and human life.

Until recently the effects of soil erosion and

soil deterioration on crop production have been masked by the increased amounts of fertilizer being applied to the land.

But the mask is slipping:

1. While consumption of commercial nitrogen doubled from 1966 to 1977, U.S. farm production increased by only 23%, 10% of which came from cropping more land.[13]

2. It is estimated that it takes five times as much fertilizer today to produce the same yields as it did in 1949.[14]

3. Yields per acre of wheat, maize, sorgum, soybeans, and potatoes have not increased since 1970.

This has happened despite the increase in chemical fertilizer usage--20% more was used in 1976 than in 1975.

If change is to come, we need a national commitment and support from our government and educational institutions to conserve our most finite resource--our soil. In 1978, land-grant institutions devoted 6,000 man-years to production efficiency projects and only 258 man-years to rural development. I think the priorities are mixed-up and would like to make a few recommendations so we can begin to turn things around:

1 Congress should require the USDA to use 50% of its research funds for small farm demonstration and research projects geared to soil and energy conservation projects.

2. A substantial amount of agricultural research funds should be ear-marked for testing new ideas and products. There is little support from our public institutions for change. There are 4,000 experts on pesticides in Iowa alone, yet I don't know any land-grant institution that teaches organic farming or the ecological approach to agriculture.

3. The testing of new products and concepts
    should be run "in the open" by the land-
    grant schools and extension services, with
    free access for the public to view the
    action, rather than being conducted in a
    shroud of secrecy.

4. I would urge the Department of Energy to
    establish a national, low-energy, demon-
    stration farm and rural life center where
    new products, innovations and concepts
    can be presented and tested. Further, I
    want the developers of these ideas to do
    the testing, openly of course, and under
    supervision, to ensure unbiased testing.

5. A major soil improvement effort is needed,
    and this will imply institutional innova-
    tions such as a permanent civilian con-
    servation corps.

There __are__ limits to growth. I am here to tell
you things are going to change whether we like it
or not. We cannot continue to use energy- and
chemical-intensive agricultural methods. Moreover,
half of the world's land available for agriculture
will be consumed by urban and industrial development by the year 2050 and population may quadruple.
There is an immediate need to support innovative
agricultural research in this country, to meet the
winds and tides of change, for the survival of
humanity, and to restore and maintain the soils on
which our food is grown.

References

1. Perelman, Michael J., <u>Farming with Petroleum</u>,
    "Environment", October 1972, Vol. 14, No. 8,
    p. 8.

2. USDA Agricultural Handbook No. 551.

3. Pimentel, David, J. Krummel, D. Gallahan,
    J. Hough, A. Merrill, I. Schreiner, P. Vittum,
    F. Koziol, E. Back, D. Yen and S. Fiance,
    <u>Benefits and Costs of Pesticide Use in U.S.</u>
    <u>Food Production</u>, "BioScience", December 1978,
    Vol. 28, No. 12, p. 772.

4. Weiman, Dave, National Farmers' Union Director
    of Research, Washington, D.C., Taken from
    speech "Land, Agriculture and People", pre-
    pared April 27, 1976

5. Hightower, Jim, <u>Hard Tomatoes, Hard Times</u>,
    Task Force on Land Grant College Complex;
    Agribusiness Accountability Project,
    Washington, D.C., 1972.

6. Brown, Lester, <u>Historic Exploitation of Crop-</u>
    <u>land</u>, "Worldwatch Paper #24 - The Worldwide
    Loss of Cropland", October, 1978, pp 7-8.

7. "To Protect Tomorrow's Food Supply, Soil
    Conservation Needs Priority Attention", Dept.
    of Agriculture, Report to the Congress by the
    Comptroller General of the U.S. - General
    Accounting Office (GAO), February 14, 1977,
    CED-77-30.

8. <u>World Crisis in Agriculture</u>, The Ambassador
    College, Agricultural Research Department,
    Ambassador College Press, Pasadena, CA, 1974.

9. James Risser, Washington Bureau Chief for the
    Des Moines Register, September 15, 1978.

10. Belanger, Jerry D., <u>Soil Fertility</u>, Countryside
    Publishing, Ltd., Waterloo, WI, 1977.

11. Zweidling, Daniel, <u>Can U.S.Farmers Kick the</u>
    <u>Petrochemical Habit</u>, "New Times", May 29, 1978,
    pp 23-65.

12. World Crisis in Agriculture, The Ambassador College, Agricultural Research Department, Ambassador College Press, Pasadena, CA, 1974, pp 19-25.

13. USDA Staff Bulletin #581.

14. Senator James Abourzek, then Chairman of Senate Judiciary Subcommittee on Deficiencies in Agricultural Research, Transcript of Hearings held in October, 1977.

Suggested Readings

1. Acres, U.S.A. Newspaper, Published Monthly, 10227 E. 61st St., Raytown, MO 64133.

2. Albrecht, William A., Soil Fertility and Animal Health, Fred Hahane Publishers, 1958.

3. Balfour, Lady Eve, The Living Soil, Faber Pub. Co., 1943.

4. Belanger, Jerome D., Soil Fertility, Country-side Publications, Ltd., 1977.

5. Berger, Samuel R., Dollar Harvest, D.C. Heath and Company, 1971.

6. Blake, Michael, Concentrated Incomplete Fertilizers, Crosby Lockwood Co., 1966.

7. Bromfield, Louis, From My Experience, Harper Pub. Co., 1955.

8. Countryside Magazine, Published monthly, 312 Portland Rd., Waterloo, WI 53594.

9. Hightower, Jim, Hard Tomatoes, Hard Times: The Failure of the Land Grant College Complex, Agribusiness Accountability Project, 1972.

10. Mother Earth News Magazine, Published monthly, 105 Stoney Mountain Rd., Hendersonville, NC 28739.

11. Pimentel, David and Krummel, John, America's Agricultural Future, "The Ecologist", Vol. 7, No. 7, pp 254-261.

12. Pimentel, David, <u>et.al.</u>, <u>Benefits and Costs of
    Pesticide Use in U.S. Food Production</u>,
    "BioScience", December 1978, Vol. 28, No. 12,
    pp 772-784.

13. Pimentel, David, <u>et.al.</u>, <u>Energy and Land Con-
    straints in Food Protein Production</u>, "Science",
    November 21, 1975, Vol. 190, pp. 754-761.

14. Pimentel, David, <u>et.al.</u>, <u>Food Production and
    the Energy Crisis</u>, "Science", November 2, 1973,
    Vol.182, pp 443-449.

15. Pimentel, David, <u>et.al.</u>, <u>Land Degradation:
    Effects on Food and Energy Resources</u>,
    "Science", October 8, 1976, Vol. 194,
    pp. 149-155.

Section B

Perspectives on Transportation

The Seneca Chief, the first canal boat through the
Erie Canal, passing through the deep cut at Lockport, New York, October 1825. Reprinted with
permission from Cadwallader D. Colden, Memoir...At
the Celebration of the Completion of the New York
Canals (New York: The Corporation of New York,
1825). Courtesy of the Canal Museum, Syracuse,
New York.

Overview

Construction of the Erie Canal illustrates the benefits of macro-engineering in the less compli-cated--and less congested--world of the early nineteenth century. Jeanne Krause shows how a relatively small investment produced profits for investors, better terms of trade for users, a healthier economy for a whole region, and an example of inter-sectoral cooperation on the grand scale. The canal was a major contributor to the "take-off stage" in national economic growth. But despite the logic of its appeal, the sums needed for the canal seemed exorbitant at the time: the whole project might have languished had it not been for the astute and persistent promotional efforts of DeWitt Clinton.

12. The Erie Canal: Macro-Engineering When the World Was Still Simple

My paper on the Erie Canal might properly carry another subtitle: rarely before have so few done so much for so many with so little. With my best apology to Sir Winston, this happened 125 years earlier than the R.A.F.'s defense of Britain.

The Erie Canal joining the Hudson River with Lake Erie was the remarkable achievement of determined men and the preeminent product of the rapid economic growth in the first decades of our republic. It was the product, too, of the close collaboration between New York officials and private entrepreneurs bound together by strong economic ties. In just seven years -- from 1819 to 1825 -- they gouged a ribbon of water 363 miles through swamp and rock, over rivers, past Indian settlements and wilderness to link New York City, "the emporium of America," with the extensive, remote frontier.[1]

This ribbon of water captured the imagination of New Yorkers in the early days of the 1800's, and naturally found its way into literature and folklore. The canal was called "The Big Ditch," "The Grand Canal," and "Clinton's Folly." Towns not on the ocean or rivers earned names such as Newport, Middleport, Lockport. For decades afterwards, when someone in a crowded room shouted, "Low bridge!" -- everyone else ducked.

---

[1]De Witt Clinton, _Memoir of DeWitt Clinton_, Reprinted by David Hosack (New York, 1829). "Ribbon of Water," Shaw, p. 87.

This ribbon did all that was expected of it, and more. DeWitt Clinton prophesied that the canal would promote national union and national defense, reduce transportation costs, deflect commerce from Canada to New York, and speed settlement of western U.S. territories.[2] This the canal did. In the process, it also helped guarantee New York's emergence as the principal city of the new nation. This extraordinary enterprise was, in Nathan Miller's phrase, "The Enterprise of a Free People." It was also a major contributor to what Walt Rostow called the "take-off" stage of national economic growth in the 1840's.[3]

The canal was the only enterprise of a people, bent on building canals in the first half of the 19th Century, that truly succeeded. Between 1815 and 1860, 4,254 miles of canals were built in 23 states for an estimated cost of $188 million. About 90 percent of that cost was financed by government bonds, half of which were ultimately held by foreigners, especially the British. The Erie Canal provided the greatest benefits and was the longest in service -- until 1910, when the current Barge Canal was built.

In part, the Erie Canal's success was due to the relative economic stability of the 1820's, when money markets functioned smoothly. And 50 years of peace were just beginning. In part, its success came because it was the beneficiary of historic and geologic coincidences, and of extraordinary luck.

New York state had a mercantilist tradition that led them to accept state aid, "regulation, and intervention in the economy." This "entrepreneurial role" of the state led to close economic ties with its citizens.

---

[2]DeWitt Clinton, "Memorial of the Citizens of New York, in favor of a canal navigation between the great western and northern lakes and the tidewaters of the Hudson." In Laws of the State of New York in Relation to the Erie Canal (Albany, 1816), pp.122-145.

[3]Walter W. Rostow, The Stages of Economic Growth: a Non-Communist Manifesto (Harvard University Press, 1960) p.38.

Two wars -- the Revolution and the War of 1812 -- had proved the military necessity of cheap transportation for defense of the frontier. For example, in 1812, a $400 cannon cost $2000 to move across New York State.

Finally the Mohawk River valley was the only major east-west break in the Appalachian chain from Georgia to Maine -- mountains which pinned the population on a thin strip of Atlantic coastline, mountains which kept abundant natural resources beyond their reach.

The bare facts were extraordinary enough. When it opened in 1825, it had 83 locks raising the water level 565 feet above the Hudson River, but lower than Lake Erie, a constant source of water. Eighteen aqueducts--marvels of the day--spanned valleys and rivers, including the Mohawk, twice. Although the canal was 363 miles long, it was only 40 feet wide and 40 feet deep. It cost $7,163,000; yet it was built by a state whose annual budget was only $700,000. For the cost estimators among you, the cost overrun was a mere one percent, and it was completed in half the anticipated time.[4]

This fragile ribbon of water was built by farmers, carpenters, surveyors, judges using rudimentary tools and some luck. DuPont blasting powder had just become available. New mechanical devices would be devised for felling trees without an axe or saw, and for grubbing 30 to 40 stumps a day. And by a twist of fate, the first segment to be built in the middle of the state passed very near an abundant supply of hydraulic cement, hitherto unknown in America.

In the early 1800's there were no engineers in America, they had to be imported. There was virtually no engineering training available anywhere until the U.S. Military Academy began in the 1820's. The Erie Canal became the school for engineers. In 1837, there were only 84 known engineers in the U.S. -- 19 from the academy and 11 from the Canal. "Graduates" of the canal would go on to build many of the other canals.

---

[4]George E. Condon, Stars in the Water: the Story of the Erie Canal (Doubleday and Co., 1974) p. 47.

Contracts were let in small segments, some as small as three-quarters of a mile. In fact, 50 men contracted for the first 58 miles in what was really a series of mini-canals, each one tested with water before settlement into the system. These contractors were mainly from adjacent areas, they furnished their own tools, hired their own labor. They used specifications drawn up by carpenters and surveyors during construction of early segments. The workers generally came from adjacent areas and, contrary to prevailing opinion, over three-quarters of them were American-born.

Most materials were found locally. "Blue mud of the meadows" provided canal lining to prevent seepage. A high grade limestone found near Medina was excellent for locks. And another variety, "meagere limestone," was the water cement mentioned earlier. Governor Clinton had read about Roman cement, and had sent a young surveyor to Europe to study canals. This enterprising young man spent his own money exploring canals, aqueducts, and experimenting with underwater cement. In 1818, he found a type easy to prepare, which hardened under water -- and he found it just miles from the first segment of canal to be constructed.

The first segment of 93 miles was built to connect the Mohawk Valley with the western counties. This choice of the middle for initial construction proved to be fortunate. Choice of the eastern section would have only duplicated an existing, if occasionally impassable, route over the Mohawk River valley. Choice of the western section would have produced a canal linking wilderness with wilderness. The middle segment, in effect, extended the potential for travel west from Utica, generated new trade in timber and grain, and passed through the salt flats on which Syracuse was built. The revenues from travel and taxes on salt and commerce were immediate proof of the canal's effectiveness--and guaranteed that the remainder would be built.

The engineering accomplishments still inspire awe. Aqueducts and culverts strode across obstacles on Romanesque arches. One embankment rose 70 feet above the Irondequoit valley.

The longest aqueduct, near Schenectady, was 1188 feet long and crossed the Mohawk River on 26 stone piers. Another, just a bit farther west, rose above the rapids at Little Falls, that had always required a portage for Mohawk River traffic. At that point, the Mohawk fell 39 feet in three-quarters of a mile. And in the middle of the rapids, the central arch of 70 feet rested on two piers driven down to bed rock.

Altogether, 27 of the locks were in the 30 miles between Schenectady and Albany. North of Albany, at the extreme eastern end of the canal route, stood Cohoes Falls -- always the most diffi-cult geologic barrier to travel. There the river fell 400 feet into the Hudson River, and 19 locks were required to bypass the falls.

In the west, the canal pierced two miles of solid rock at Lockport, where a double flight of five locks carried the canal 66 feet over a rock ridge. Those locks required 50,000 feet of facing stone, were each 90 feet long and 15 feet wide, and cost $1,000 per foot of rise. Floors of the locks rested on timbers one-foot thick, which, in turn, were covered by two layers of well-jointed planks three and two inches thick. All timbers were thoroughly grouted and clamped in iron. And Lockport did not exist until July of 1821, when construction began.

For those investors among you, several years before 1825, revenues from the still incomplete canal exceeded interest costs; from the time the canal opened, tolls were double the debt interest. Moreover, surplus revenues from the canal, which the state reinvested, themselves earned interest. Seven years after its completion, revenues had surpassed the entire cost of construction. Tolls were abolished in 1883. By then, all revenues had totalled $122 million. All construction, super-vision and maintenance costs had come to only $80 million. Construction, that is, of two canals- for New York decided to double the size of the canal just 10 years after it opened.

If you actually had been an investor in canal fund bonds, you or your heirs would have reaped another, unanticipated benefit. The commissioners, unable to retire the debt early, sought alternative

investments for surplus revenues. Among their
choices: newly-formed companies building what we
know as the New York Central railroad system.

Funds for construction of the canal and revenues from its operation contributed to the economic
development of regions through which it passed.
Construction was paid for by engineers, using drafts
on local banks. As revenues from passengers, salt,
and goods poured in, they too were deposited in
local banks; in 1826, just one year after opening
the canal, the canal commissioners, unable to
retire the debt -- then selling about 25 percent
above par -- began depositing millions of dollars
in New York City and banks along the route, at
5 percent interest. After 1831, the canal fund, in
effect, became a development bank, as commissioners
also loaned money to the banks. By 1836, they had
loaned $3.6 million to 52 institutions. In the
1837 panic, the commissioners loaned $2.5 million in
canal stock to seven New York City banks.

As one might imagine, shipping costs were
drastically reduced by the canal, better for commerce than mud-clogged roads and unpredictable,
flood-swept rivers. Before the canal, it took
three weeks to move one ton of goods from Buffalo
to New York -- at a cost of $100. The canal cut
time to nine days and the cost to $10. More remark-
able, the 90 percent reduction in costs wrought by
the canal was more drastic than any subsequent
differential brought by the railroads, and it
allowed Americans for the first time to make effective use of the great interior and to establish the
national market on which later industrial development could be based.

Population, of course, grew rapidly. To cite
but two cities: in 1820 Rochester had only 1500
people, by 1850 over 36,000. Between 1820 and
1840, the number of people employed in manufacturing almost trebled, and those engaged in commerce and navigation rose tenfold. By 1835
Rochester had become the milling center of the
state, its 21 mills using 96 stones. The city
served the farmers who had planted wheat fields 40
miles deep on either side of the canal as it passed
through Genessee valley. In 1830, 13 percent of
all tolls on the canal were paid in Rochester.

Buffalo swelled from a hamlet of 2000 in 1820, to over 20,000 in 1835 -- the year $1 million worth of buildings went up. By 1850, the population had doubled again. The same occurred in other cities along the route and those that were to be founded in the west.

Land values were naturally affected. The Federalists' interest in the canal followed their interest in land. Many state officials, legislators, and canal commissioners promoting the canal, and later responsible for building it, were speculators in land. They were not disappointed. In one wilderness county crossed by the ultimate route, land values quadrupled three times: after the canal land survey was authorized in 1808, after the canal act of 1812, and again after the final act passed in 1817. They almost doubled again by 1819 when the first 83 miles opened nearby. What happened to those land values in the years in between, speculators know only too well.[5]

Politically, the canal had as great an impact as it had economically -- and was as subject to accident. As long as the idea was in the hands of visionaries, it made little headway in Albany or Washington, where early presidents were advocating using federal surpluses for internal improvements. Visionaries who were, by the way, wealthy Federalists owning large tracts of land and western New York legislators -- those who stood to gain the most from a canal.

The idea made little real headway for a generation, since the 1790's when an earlier attempt at a canal foundered because of bad management and under-capitalization, and its two miles actually built had wooden locks that were rotten almost before the canal opened. That abortive attempt did influence the New Yorkers' thinking about internal improvements. Failure to improve rivers strengthened the case for an independent canal, and financial failure of a private company led promoters to favor a public venture.[6]

---

[5]Jeanne Krause, unpublished master's thesis, "Speculation in Land Along the Route of the Erie Canal by New York State Officials 1793-1822" (University of Chicago, 1975).

Then, almost simultaneously, the idea of a canal got the support of New York City merchants and DeWitt Clinton -- a great, imaginative leader who organized the coalition that voted in the canal. Not a machine politician, Clinton had been drummed out of office as mayor of New York. Twice. It was easy in those days. After the second time, in 1815, he took up the issue of a canal, mobilized opinion throughout the state, and rose to the state house on a wave of sentiment for it. The coalition which passed the final act in 1817 was made up of Federalists and his splinter "Clintonians." Only two Federalists voted against the final survey enacted in 1816, and all voted for the 1817 act.

Clinton was masterful in overcoming opposition. To legislators and landowners from the Hudson River, he promised that New York City's population would grow at least as fast as the supply of competing goods from western New York. It did. To northern and southern counties not crossed by the Erie Canal, he promised a Champlain canal to link New York City with the St. Lawrence River and lateral canals to the Erie. They were built. In the process of forging a coalition of Federalists and western legislators, Clinton set in motion a process that would, even before the canal opened, move New York politics out of the hands of "the few the rich, and the well-born" into more democratic patterns. A move -- in the most populous state -- some scholars say was essential to the election of Andrew Jackson.[7]

Two other accidents are worthy of note. In the final days of legislative argument in 1817, people were still uneasy about the financial success of the canal until Martin Van Buren -- an early opponent of the canal -- tied his political future to the canal. We all know what his political future was. In the final days of argument, Van Buren proposed that canal fund bonds be backed by the credit of New York State. At that

---

[6]Nathan Miller, "Private Enterprise in Inland Navigation," in New York History, Vol XXXI, No. 4, October 1950, pp. 398-413.

[7]"The Few, the Rich, and the Well-Born," is chapter one in Fox.

time, private capital was in chronic short supply
in America, and no state, including New York, was
willing to raise taxes to build canals. They,
therefore, had to rely on borrowing -- made much
easier with the financial backing of the state.
That proposal proved crucial to the act's passage.

In those days, the council of revision -- not
the governor -- held power of veto, and the vote on
the 1817 act was three to two against. Until former governor, Daniel Tompkins -- then vice president of the United States -- walked into the
chamber and joined the debate. His opinion, he
said, was that renewed hostilities with Great
Britain were inevitable and that funds should be
saved for war. Wherupon Chancellor Kent responded
that he preferred a canal to war and voted yes.
The act carried three to two.

The effort related to the canal contained
certain ironies. In the early decades of the new
republic, a tension existed between New England and
the government in Washington. Federal jurisdiction
over interstate commerce was not yet firmly established. The Erie Canal did more to force the government into regulation of that commerce than anything prior to the railroads. The New York
Federalists worried about a possible shift of the
national center of gravity away from New York to
the Potomac and, under Jefferson, away from commerce and industry to agriculture. However, being
in favor of a strong central government, they
naturally turned to Washington for support and
funds. Refused by three presidents and as many
Congresses, New York State built the canal itself
--thereby delaying the shift of commercial and
financial power further south, but also, in a very
real sense, tying together a fragmented nation.

Now, to us, it seems inevitable that the
canal would have been built and its success assured.
To citizens of New York State in the early 1800's,
the scale of the endeavor appeared phenomenal, the
cost exhorbitant. Technology was not yet adequate.
Native engineers did not exist. To fully grasp
what this enterprise meant to those citizens, let
me suggest a comparable enterprise today: it would
be as if Americans were to build a rapid transit
system to the moon, without physicists or computers
at a cost of $5 trillion -- to be borrowed mainly

from the Russians and Chinese.

Would we calculate the costs and the benefits accurately? Would we accept the challenge?

Bibliography

Fox, Dixon Ryan, The Decline of Aristocracy in the Politics of New York (Columbia University Press, 1918).

Goodrich, Carter, editor, Canals and American Economic Development (Columbia University Press, 1961). Especially the chapters by Julius Rubin, "An Innovating Public Improvement: The Erie Canal," and by Harvey H. Segal, "Cycles of Canal Construction."

Miller, Nathan, The Enterprise of a Free People: Aspects of Economic Development in New York State During the Canal Period 1792-1838. (Cornell University Press, 1962).

Shaw, Ronald E., Erie Water West: a History of the Erie Canal 1792-1854 (University of Kentucky Press, 1966).

Whitford, Noble E., History of the Canal System of the State of New York, a Supplement to the Annual Report of the State Engineer and Surveyor for the Fiscal Year Ending September 30, 1905, two vols., (Albany).

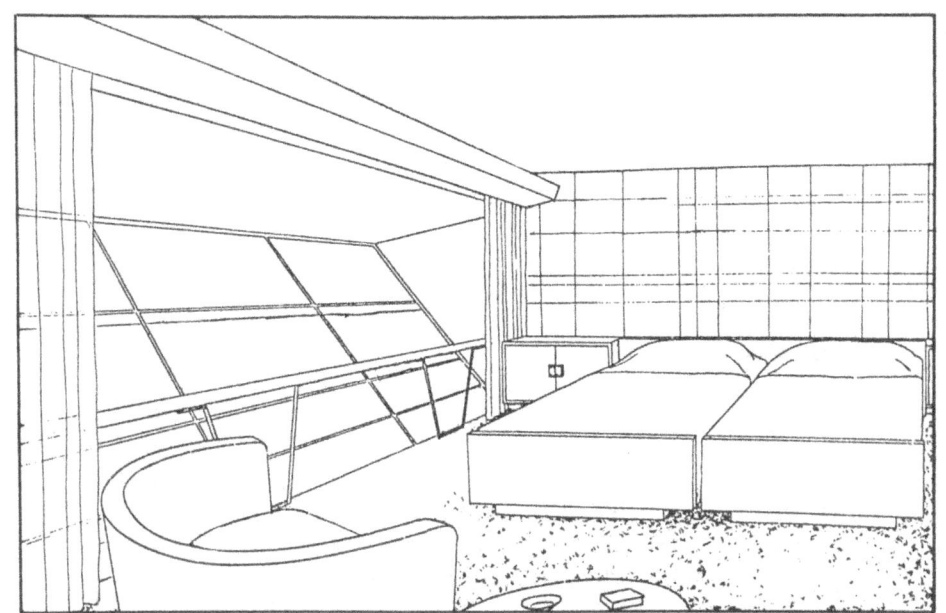

Conceptual drawings by Francis Morse of a "Deluxe Cabin" (above) and a "Sky Room" (below) in a "new generation" airship.

Overview

A new generation of giant lighter-than-air craft could provide a valuable type of air transport and find a viable economic niche in the spectrum of the transport industry. Francis T. Morse explains that structural efficiency can be enhanced by an increase in size to a point where an airship's ratio of useful lift to gross lift exceeds that of its heavier-than-air counterpart. Airships could utilize engines and fuels not suitable for heavier-than-air craft, one such possibility being a turbine system using liquid hydrogen for fuel. Among potential benefits of a new generation of lighter-than-air craft would be low-cost bulk transport and reduced air and noise pollution.

Francis Morse

13. The Future of Large Airships

The demands of long-range transportation have spawned an era of gigantism in both surface craft and aircraft. Notwithstanding certain disadvantages, the trend should generally not be unfavorably regarded, although it must be noted that its chief beneficiary is <u>operational economy</u> rather than <u>structural efficiency</u>. However, in the case of lighter-than-air craft, structural efficiency-- a prime criterion for successful design -- can indeed be enhanced by increase in size, to a point where an airship's ratio of useful lift to gross lift far exceeds that of its heavier-than-air counterpart. This interesting statistic may prove compelling in spurring the renewed development of large lighter-than-air vehicles. But another factor will increasingly mitigate in favor of the airship -- the impending depletion of petroleum reserves to a point where alternative fuels must be introduced. Again, the giant airship casts its shadow upon the transportation picture.

It would be misleading to picture airships as playing the same role in aviation as airplanes; the inherent characteristics of each type are too divergent. The modern airliner has realized the role of the airplane in superlative fashion. But it has left gaps in the transportation spectrum which remind us of the potential value of the airship's role -- a long neglected one! If an airliner's principal asset is its speed, an airship's are surely comfort and luxury. The cruising speed of a passenger airship may never exceed 125 mph., but its potential for restful and luxurious travel is almost unbounded. Similarly, in the realm of air cargo, while the jet freighter can ferry big

loads across the oceans overnight, the mean density
of such cargo must be high because cargo space is
limited. In an airship, weight - not space - is
the only limiting factor, and consequently low
density cargo can be transported effectively on
intercontinental routes; if not overnight, in 36
hours at most, and possibly at lower DOC's per
ton-mile.

With the above considerations in mind, let's
examine the features and performance which will
characterize a new generation of rigid airships.
The hull of an airship is intrinsically a highly
efficient space frame. With a fineness ratio of
between 5 and 6 to 1, it is also efficient aerodynamically. Its drag is almost wholly skin
friction. Past airships have not always been built
with efficient hull structure, nor have they always
been ideal aerodynamically - but this has been
because designers failed to maximize their opportunities. Unlike heavier-than-air craft, the airship benefits greatly from increased size. This is
because of the cube-square law which influences the
ratio of dead weight to gross lift. I.e., since
the hull volume (lift) increases as the cube of a
linear dimension and the surface area (a fair
measure of structural weight) as only the square of
a dimension, the larger the ship, the smaller
becomes weight/lift ratio. This fact should become
a strong incentive to larger sizes in future airships. The Hindenburg and its sister ship represented the largest LTA vehicles ever built - each
having hydrogen capacities of 7,000,000 cu. ft.
with corresponding gross lifts of 500,000 pounds.
Since existing hangars will, as a practical exped-
ient, probably be used for the next rigid airship,
a size limitation of some 12,500,000 cubic feet is
indicated. Inflated with noninflammable helium its
gross lift will average nearly 800,000 pounds.
(more in winter, less in summer). It is sheer
coincidence that this equals the takeoff weight of
a Boeing 747, but it provides a welcome basis for
comparison between the two types. Like the airliner, the airship will carry some 400 passengers.
Unlike the jumbo jet, the airship will have deck
space totalling some 40,000 square feet -- approximately 100 square feet per passenger, or ten
times that of the airliner. The accommodations
will, in fact, much more resemble those of a surface ship than an airplane. There will be state-

rooms, promenades, lounges, cinemas, saunas, gymnasiums and other novel amenities.

Structurally, it appears that for the size hull in question, the optimum structural arrangement should pretty much follow past practice, i.e., with cantilever or wire-braced transverse frames and circumferentially spaced longitudinal girders built up from seamless tubing of high-strength aluminum alloys. Resistance to shear is achieved by a diagonal network of steel wiring linking the girder joints. Over this metallic skeleton will be stretched a taut outer cover of aluminized Kevlar fabric. Within the hull, an array of 17 to 18 Tedlar-lined helium cells, each equipped with a set of maneuvering and automatic over pressure valves for the discharge of helium. All living quarters-- passenger accommodations, crew quarters, control bridge, etc., will likewise be contained within the hull space so that the only external appendages will be the horizontal and vertical control surfaces, or, as they are called in LTA terminology, the "fins."

Other than the big jump in size over past prototypes, the most important innovation marking a new generation of airships will be its propulsion system. The last Zeppelins were driven by diesel engines mounted in "pods" slung from the sides of the hull. Owing to the rather high specific weights of these power plants, this may have been the only practical arrangement at the time, but it was hardly an efficient one. It was mentioned above that the drag of a well-streamlined hull at large Reynolds numbers consists purely of skin friction. Consequently, an airship drags behind it in flight a wake in the form of an invisible cylinder of air of about half the hull's mean diameter.

On the other hand, each of the propellers driving the airship generates a slipstream, moving in the opposite direction. Both wake and slip-stream contain large amounts of wasted kinetic energy which can be chalked up to drag in the one case and low propulsive efficiency in the other. By locating the propellers in the wake, these two sources of lost energy can be made to cancel each other out! Propulsive efficiency will improve dramatically. Fuel consumption will drop sharply--

and payload correspondingly boosted. The huge hull
of an airship represents the ideal application of
stern propulsion, and along with efficiencies
approaching 100% will come other benefits. For
example, by incorporating very large diameter, slow-
turning counter-rotating propellers the sound level
can be reduced by as much as 80-90%. The airship
of tomorrow will be the first truly silent ship of
the skies - a welcome new-comer to an age plagued
by noise pollution. Also, by placing the propellers
aft of the fins, the flow of air over the roots
of the fins will be improved, thus permitting a
reduction in overall fin area.

What type of engines will drive these sixty
foot diameter propellers? What energy source will
supply them? Here the airship designer will enjoy
a flexibility of choice largely denied to his
heavier-than-air colleague. For quite obvious
reasons, today's jetliner can be powered only by
petroleum fueled turbofan engines - miracles of
compactness and low specific weight, but noisy and
limited by low Brayton cycle thermal efficiency.

A conservative approach to any new airship
power plant might be to incorporate an improved
diesel engine, with a s.f.c. fully twenty percent
lower than that of the best turbofan engine available. This would be equivalent to adding 20,000
pounds to the payload of a 747 jumbo jet on a
transatlantic hop--at substantially lower costs per
gallon of fuel. But an equally feasible and
potentially far more advantageous solution is to
make use of the Rankine cycle for airship power
plants: a modernized version of the steam turbine
energized by liquid hydrogen as the fuel. LH_2 has
a heating value almost three times that of hydrocarbon fuels. What's more, it derives from a
renewable source--water. Its exhaust products
are non-polluting H_2O. And liquid hydrogen is as
safe to store and handle as any other liquid fuel.
However, since LH_2 boils at 36.7° Rankine, it is
a cryogenic fuel and this will entail a fuel
system of greater weight and complexity than a
conventional one. Furthermore, since the density
of LH_2 is low (only 4.43 lb/ft^3) the space which
the fuel tanks must occupy is large (a real
stumbling block for airplanes, but no great handi-
cap for airships). For airships, at least, the
advantages of LH_2 far outweigh its disadvantages

and it is uniquely suited for LTA propulsion. In
the airship under discussion, the boilers would be
located within the lower portion of the hull forward of the fins and the turbines, with reduction
gearing, adjacent to the stern-positioned dual
rotation propellers. And the condensers--normally
the heaviest component of a Rankine system--would
be the vast network of tubing comprising the
structural girders of the hull itself. Here,
rejected heat would be absorbed by the ship's
helium to produce additional lift--a real bonus
to boot. The effect that all this will have on the
payload of a large airship is illustrated in
Figure 1.

Once the rigid airship has re-entered the
transportation field, its evolution should be
rapid. Its uses will undoubtedly be varied,
ranging from long-range cargo hauling and scheduled
transoceanic passenger service to more exotic uses
such as luxury cruise vessels, oceanographic
research applications and disaster relief vehicles.
Great new hangars will be built, permitting the
launching of truly mammoth airships. A hangar two
thousand feet in length would suffice for the construction of the hundred million cubic foot behe-
moth outlined in Table 1. An airship of this size
would benefit greatly from a nuclear power plant,
as the table indicates. Even the high weight of
the reactor shielding would become a minor item in
an airship of such proportions, and the added cost
of nuclear operation would be easily absorbed by
the greater payload. It should be noted that the
linear dimensions of this 3,000 ton airship would
be only twice those of the 12.5 million cubic foot
ship of just one-eighth its volume and lift. With
a hull of such large capacity, new structural
arrangements other than the traditional space
frame would be called for. Transverse frames and
longitudinal members would be merged into an
integrated trusswork surrounding the helium cells
and supporting the outer cover. Shear and radial
wiring would be eliminated (except for bulkhead
wiring to contain the gas cells.)

Quite obviously, a three-dimensional space
frame of such size and complexity will present a
major challenge to the designer. Currently, large
clear two-dimensional spans are accommodated by the
standard flat octet space truss (airport hangars,

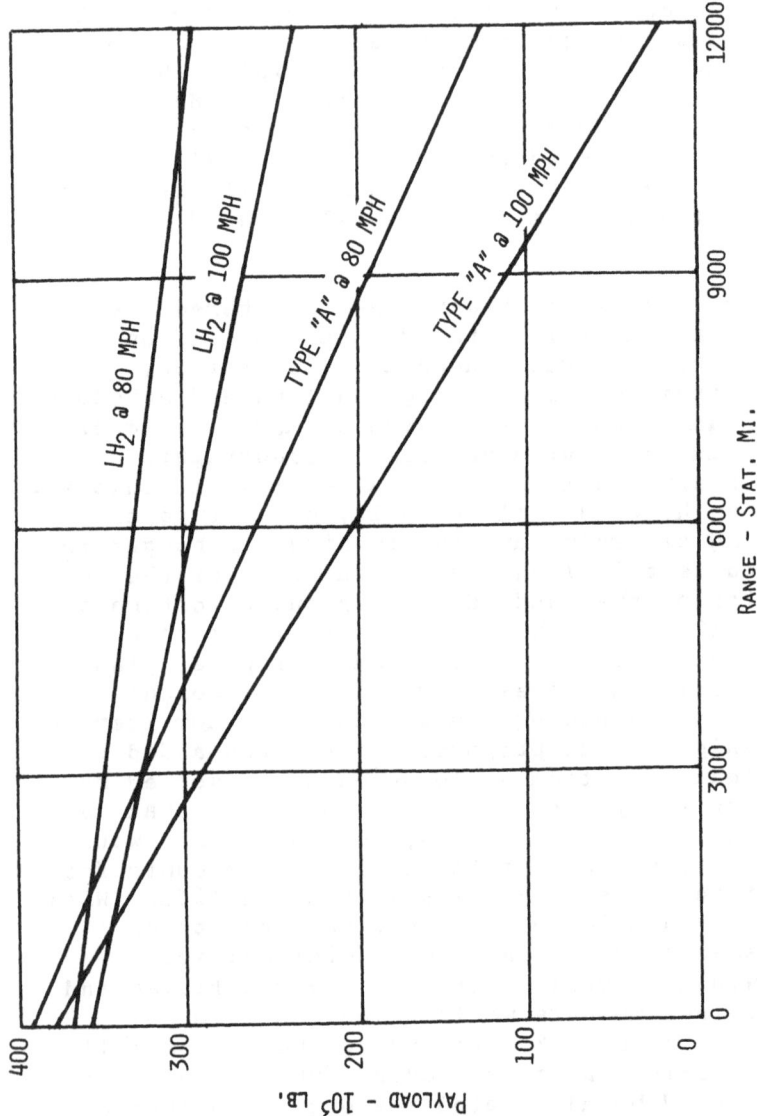

Figure 1. Payload U.S. range for 12.5 million cu. ft. cargo airship using LH₂ or Type "A" jet fuel.

Table 1. Principal characteristics of a 100,000,000 cubic foot rigid airship.

Item	Conventional Power Plant	Nuclear Power Plant
Overall Length, Feet	1,900	1,900
Maximum Diameter, Feet	333	333
Fineness Ration (L/D)	5.7	5.7
Surface Area, Sq. Ft.	1,530,000	1,530,000
Maximum Gas Volume, Cu. Ft.	100,000,000	100,000,000
Volume of Displaced Air, Cu. Ft.	106,000,000	106,000,000
Gross Lift @ 95% Helium Inflation, tons	3,000	3,000
Useful Lift, Tons	2,150	2,000
Payload @ 6,000 Mi. Range, Tons	1,750	1,850
Payload @ 12,000 Mi. Range, Tons	1,500	1,850
Number of Gas Cells	20	20
Number of Engines	8	8
Total Horsepower	32,000	32,000
Maximum Speed, MPH	120	120
Cruising Speed, MPH	108	108
Number of Personnel (Operating)	60	66

sports arenas and the like). Where a doubly curved surface--e.g. the envelope of an airship hull-- is concerned, the problem becomes far more diffi- cult. Fortunately, studies in the geometry of folded plate systems by Prof R. Resch of Boston University promise to simplify the procedure of airship hull design and at the same time to offer a feasible method of repetitive element hull construction. Using computer assisted techniques, Resch and Christiansen have defined a series of versatile structural systems which are capable of producing an infinite variety of enclosures (Figure 2). An arrangement of a series of these structures would be eminently suitable for the envelope of a very large airship hull. As well as being both lightweight and strong, the modular pattern would lend itself to economical mass pro- duction techniques. Their computer program allows the specification of any surface of revolution. It will proceed to construct a truss tangent to that surface, the depth of which may be specified, and will output a control tape for the description of all the members of the given structure.

Within this gigantic hull would be disposed luxury hotel accommodations for 3,500 passengers; promenade decks longer than the QE2; hangar space for half a dozen shuttle planes or helicopters; cargo holds totalling a million cubic feet or more. Truly, it could be termed the ultimate airship.

A number of structural and aerodynamic ad- vances resulting from enlarged displacement have been described in the preceding paragraphs. A further aerodynamic effect deserves mention here. Airships are noteworthy for their smooth ride -- no case of airsickness in a rigid airship has ever been recorded -- and the larger the airship, the steadier its flight path. This is expressed graphically in Figure 3 which indicates the effect of size increase on the ratio gross lift to sur- face area in pounds per square foot. At a ratio of 2 lb/sq.ft. of surface area, all motion due to normal atmospheric turbulence is effectively suppressed. This corresponds to a hull length of approximately one thousand feet.

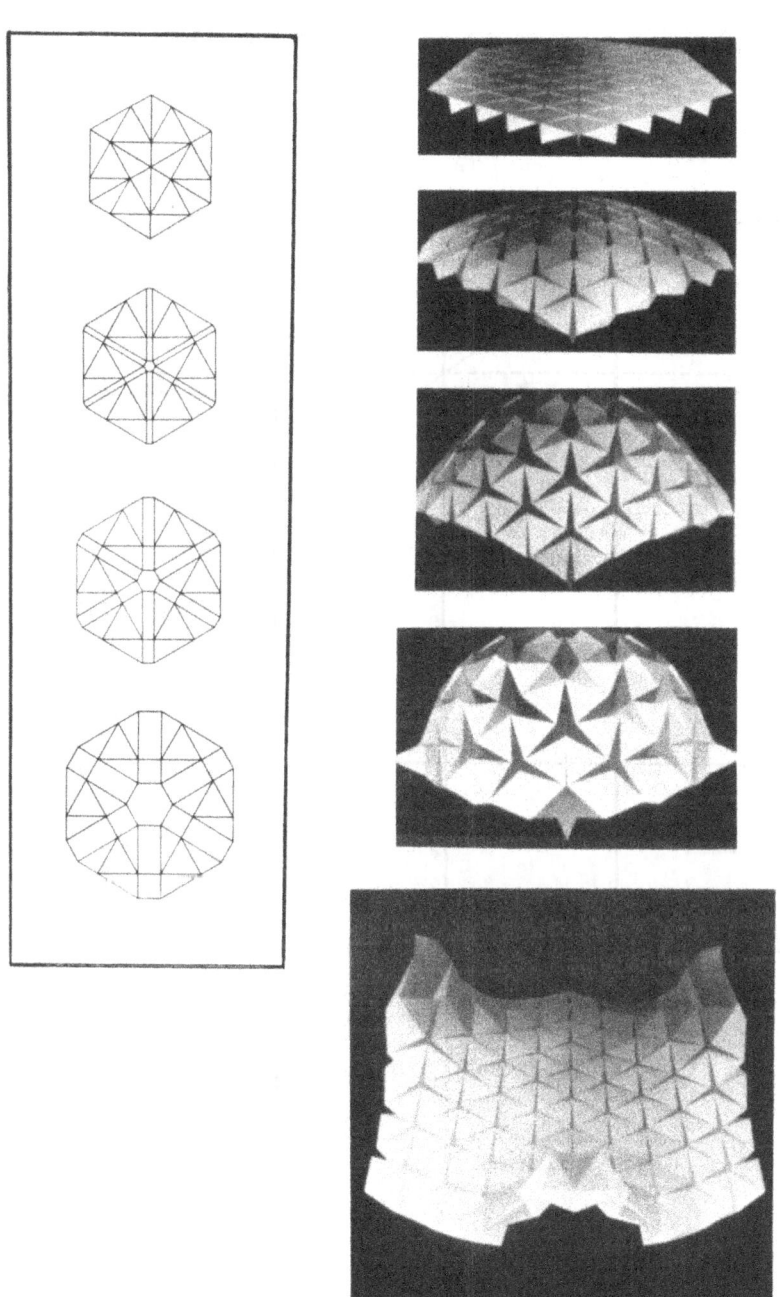

Figure 2. Computer-generated structural systems for various types of enclosures.

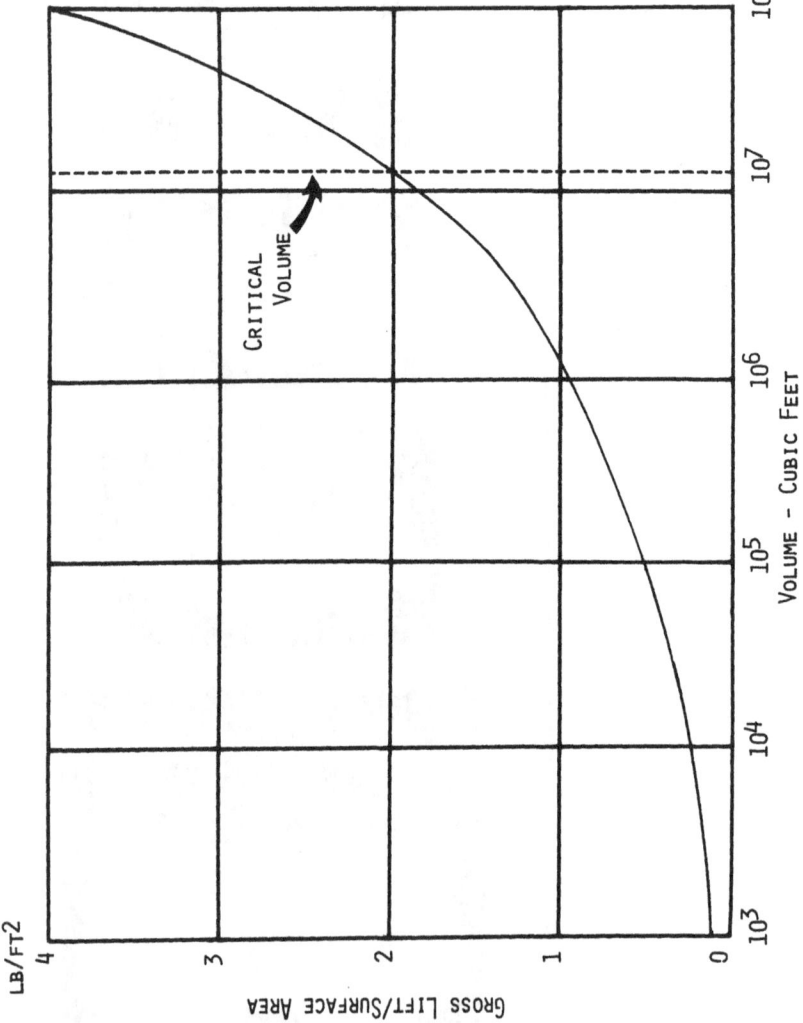

Figure 3. Effect of size increase on ratio of gross lift to surface area.

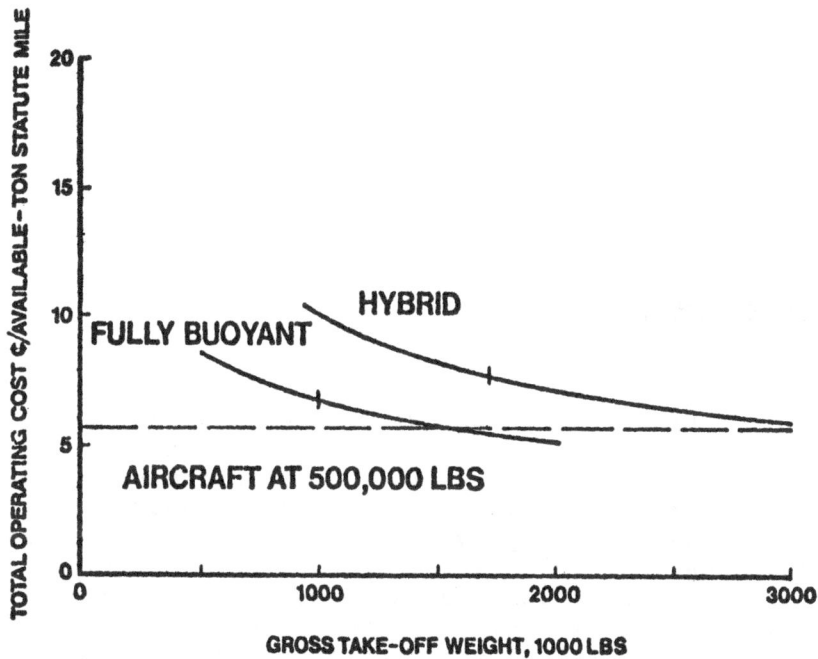

Figure 4. Effect of take-off weight on operating costs for three types of aircraft.

A detailed discussion of airship operating costs is outside the scope of this paper. However, a comprehensive indication of the effect of size on direct operating costs for airships vs. airplanes is shown in Figure 4. Once again, the scale advantage is apparent, mainly because of the decreasing empty weight fraction and improved aerodynamic efficiency with volume growth. In today's competitive society, this alone should be reason enough for the energetic development of new and larger airship types.

Overview

A number of interesting alternatives exist for improved high-speed travel between distant points on the globe. These include electromagnetic levitation trains in vacuum tubes, laser-powered aircraft, hypersonic flight, suborbital rocket-powered vehicles, and orbital transport. J. Peter Vajk stresses that technological, economic, and social acceptance questions still remain to be answered for all of these alternative transport systems. Perhaps the most likely candidate to survive the multi-criteria assessment of these systems is the orbital transport.

J. Peter Vajk

14. Global Transport Systems for the Turn of the Century

The desire to explore the world in which we live appears to be a universal instinct of every living creature. In human consciousness, this urge is experienced as a sometimes painful schizophrenia between the "here and now" and the "there and then." Although it is still possible to find people who have lived their entire lives within just a few kilometers of the house in which they were born, most people on earth today have traveled far more extensively than would be required for physiological survival. In the more affluent nations of the world, we often travel great distances, sometimes in the course of our jobs, sometimes for pleasure as tourists. We casually accept the idea of vacation trips to visit the natural wonders of the world; yet the appreciation we have for natural landscapes, in and of themselves, is a very recent value in Western civilization, having originated with the English Lake poets about the end of the eighteenth century.[1]

Today, extensive networks of highways, railroads, shipping lines, and airlines make travel to far away places far easier than could have been imagined at the beginning of this century. Supplemented by sophisticated networks of worldwide electronic communications, this same transportation system has brought about a progressive increase in the economic, material, political, and cultural interdependence of the nations of the world. Within the last generation or two, the "global village" has become a reality. As our interdependence continues to grow, so will our need for still more efficient transportation systems for long-range travel.

This paper was originally conceived as an assessment of one particular technological extension of the global transportation system, the development and commercial implementation of hypersonic flight. In order to assess the viability of hypersonic flight around the turn of the century, however, it was necessary to consider various alternative extensions of the global transportation system, and it seemed more natural and useful to develop a preliminary comparative assessment of five potential new transport systems which could, in principle, be implemented by the turn of the century.

The Need for Global Transportation

What purposes would be served by improving the global transportation system beyond the present technology represented by such aircraft as the Boeing 747 and the Concorde SST? Why should anyone want to travel any faster than these sleek mechanical birds can fly? Is hypersonic flight the next logical step in the evolution of the global transportation system?

Let's consider the uses of the global airline system as it stands today. Tourism, of course, is an important function of this system, providing a great deal of personal satisfaction, pleasure, and enrichment to many. Air travel has enhanced our knowledge and appreciation of the biosphere and of the diversity of human culture and experience. For most inhabitants of the New World, it has made it possible to reestablish our appreciation for our own ancestral and cultural roots. The recent breakthroughs in trans-Atlantic fares triggered by Sir Freddy Laker have already expanded the personal experiences and educations of countless students who can now afford to visit Europe. The human value of alleviating the pain of wanderlust is not to be scorned, either.

But on a routine, round-the-year basis, the backbone of the airline industry's revenues comes from the business and government sectors. Air freight shipment of high priority cargoes, air mail service, and air travel by business executives, sales representatives, and government officials are of critical importance to the day-to-day operation of the interdependent global economy. In the last few years, this has become especially visible in the form of thousands of American managers and engineers who are helping some of the major oil-producing nations of the world (Saudi Arabia springs to mind here) to develop their resources and to bootstrap themselves into the modern age.

The business executive who can now return from Europe aboard a Concorde the same afternoon of his (or her) departure from New York, carrying signed contracts worth several million dollars, has become a routine element of international trade. But the value of such commercial uses is not exclusive to the industrialized, affluent nations. Such travel is enormously valuable to business and industry in the Third World as well, since it has given them access to trade fairs and global markets which were scarcely aware of the existence of many companies as recently as ten or twenty years ago. In the absence of personal business contacts, Third World companies could scarcely participate (let alone compete) in the international marketplace.

"Shuttle diplomacy" is a phrase coined in the 1970's to describe the effective use of rapid air transportation between the political centers of the world. While shuttle diplomacy was most dramatic in the Middle East where air travel times are as little as an hour or two, the current transition in U.S.--China relations (whatever one's opinion of the political wisdom of that transition) could not have been effected nearly as rapidly in the absence of the global air transportation system.

The global transportation network has also facilitated major international development programs aimed at alleviation of human misery in the most impoverished parts of the world. Disaster relief, famine relief, and evacuation of refugees have occasionally relied on massive airlifts. Air transportation on a more routine basis has brought expertise, equipment, and managerial skills into underdeveloped areas, while permitting students, scholars, business executives, and government officials from Third World countries to study, consult, and learn abroad. Although the rapidly expanding electronic communications networks have made immeasurable contributions to this process as well, the electronic media are not--and never will

Table 1. Some major cities in the global
transportation network.

Berlin	Lisbon	Peking
Bombay	London	Rio de Janeiro
Buenos Aires	Los Angeles	Rome
Cairo	Manila	San Francisco
Calcutta	Mexico City	Santiago
Capetown	Miami	Seattle
Caracas	Montreal	Shanghai
Chicago	Moscow	Singapore
Colon	Natal	Stockholm
Dakar	New Orleans	Sydney
Hong Kong	New York	Tokyo
Honolulu	Novosibirsk	Warsaw
Istanbul	Paris	Washington
Khabarovsk		

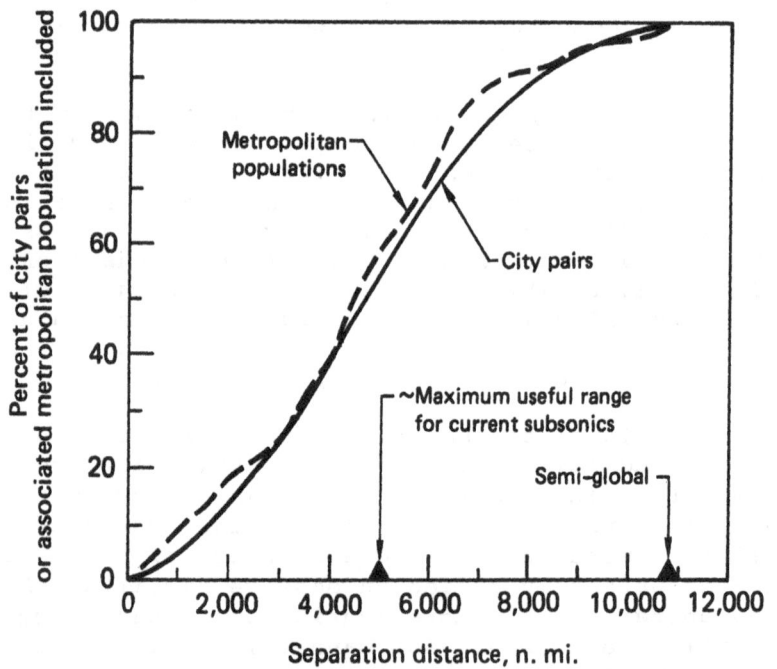

Figure 1. Sample distribution of inter-city dis-
tances in a globally interdependent economic system.
Adapted from Robert Salkeld, "Global Rocket Trans-
ports: Changing Perspectives," in The Second Fifteen
Years in Space, Vol. 31, Science and Technology,
American Astronautical Society, San Diego, 1973.

be--a complete substitute for personal face-to-face discussions and for on-site experience.

While present day jet aircraft are far more reliable than their piston-engine predecessors, mechanical failures and breakdowns do occur on occasion. Should this happen to an airplane in Boston or London or Frankfurt, repair facilities, personnel, and replacement parts are readily available on the field, and the flight may be delayed an hour or so. But if a jumbo jet has a mechanical failure in Montevideo or Nairobi or Karachi, chances are high that the replacement parts must be flown in from the United States or Europe, in some cases with tools and mechanics as well. The delay before that aircraft can return to service may then be 24 to 72 hours, because of the lengthy travel time for the parts, tools, and repair personnel. Where alternate flights are infrequent or have few empty seats, most stranded passengers in this situation would also be delayed up to three days behind their schedules, and the airline's entire timetable may be severely disrupted. Economic losses can total far more than the direct inconvenience to a few hundred passengers.

All of the present uses of the airline transportation network around the globe would clearly benefit from improvements in its effectiveness and capabilities, especially as longer range trips become more and more common, up to trips half-way around the globe.

The maximum practical non-stop range for subsonic jet transport aircraft in use today is about 5,000 nautical miles (5,750 statute miles or 9,200 kilometers). As shown in Figure 1, nearly half of the 40 major airline hub cities listed in Table 1 are separated from each other by more than this distance. The 40 cities listed are not the largest in population, but were included on the basis of their present character as regional hubs for airline transportation, size of population, and political, economic, and cultural importance. The distances between cities for each of the 780 different pairs of cities were computed as great circle distances; because of lack of navigational facilities or because of political necessities of flying around certain territories, actual air route distances are in all cases somewhat greater than these great circle distances.[2]

Figure 2 shows the travel times associated with long-distance travel by present day subsonic jets and by the Concorde supersonic transport. Depending on the traveler's personal decisions (or on available flight schedules), traveling half-way around the world today by subsonic jet provides a choice of two different ordeals by endurance. Either one flies straight through, with one or more refueling stops or flight connections, for a twenty-four hour or longer siege; or one breaks the trip with an overnight stop, which almost invariably involves clearing customs and immigration, ground transportation to the hotel district, partial unpacking of luggage, repacking after a poor night's sleep, transportation back to the airport, all of this compounded by the possibilities of travel time of 35 or more hours. The situation is somewhat improved for supersonic transport, with the possibility of getting half-way around the globe in less than ten hours if flight connection times or refueling stops are short and uneventful. Breaking the trip with an overnight stop results in a total travel time of a bit more than twenty hours. Note that the times shown in Figure 2 are optimal--they assume great circle routes can be flown; that headwinds are moderate; that connection times or refueling times are ideal; and that flights are available just when the traveler wishes to go.

Figure 2 also shows travel times for a Mach 8 hypersonic transport and for a global orbital rocket, but I want to defer discussion of these possibilities until later. Note, however, that the hypersonic transport would allow semi-global distances to be covered in a total travel time of as little as five hours, a time which is commonplace today for transcontinental flights in North America today by subsonic jets. Travelers on such flights do experience a significant level of fatigue and stress, but the experience is seldom perceived as an ordeal of endurance.

Global Transport Alternatives

What are the possible options for extending the global transportation network toward more

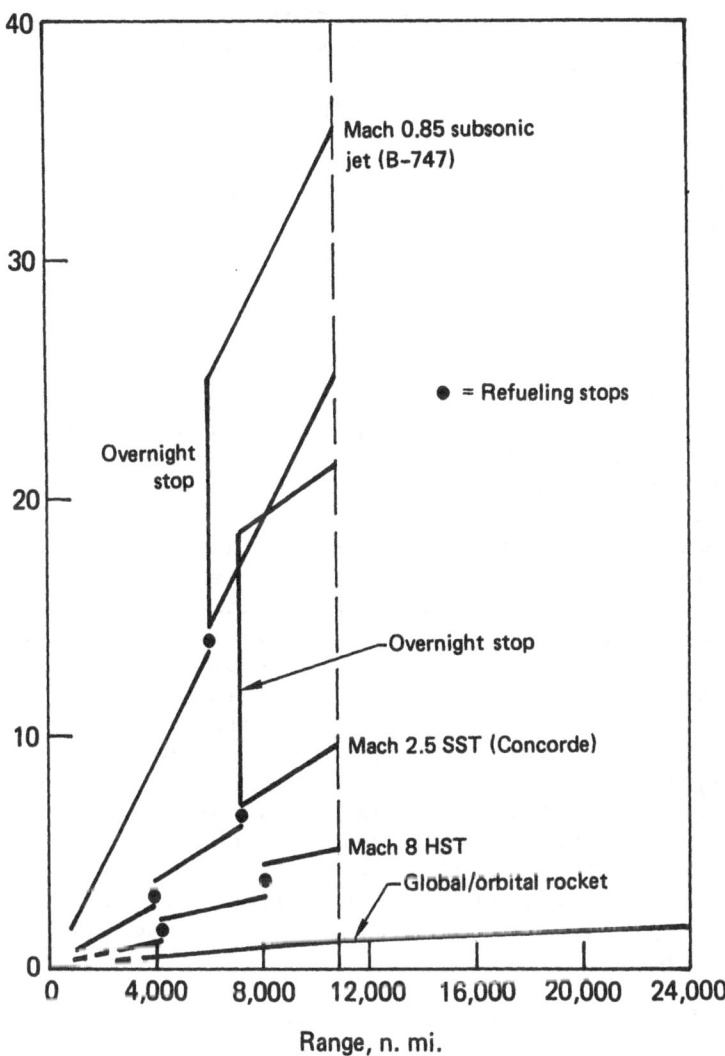

Figure 2. Travel time as a function of distance by current and projected flight technologies. Adapted from Robert Salkeld, "Global Rocket Transports: Changing Perspectives," in <u>The Second Fifteen Years in Space</u>, Vol. 31, <u>Science and Technology</u>, American Astronautical Society, San Diego, 1973.

efficient and rapid transportation between two cities more than 5,000 nautical miles apart? Obviously, transportation paths between two cities on the surface of the earth may lie beneath the surface of the planet, on the surface, in the atmosphere, or through the atmosphere into space and back. Before discussing new options, I want to indicate briefly some of the reasons why surface transportation cannot be expected to contribute to a major extension of the global transportation system.

Land transportation requires smooth roadways if it is to be fast and safe. On natural surfaces which are reasonably smooth (such as snow and sand) sliding vehicles have been used since prehistoric times, but these are very inefficient because of friction losses. Tracked vehicles have proven very useful on uneven surfaces, but their mechanical complexity restricts their speed and reliability. The biological solution of articulated legs has been highly successful in the animal kingdom for hundreds of millions of years, but only because biological computers are sufficiently compact to perform the complex calculations needed to control the motion and placement of the legs on irregular surfaces and to interpret the kinesthetic feedback required for even stress distribution among the legs. Aside from the mechanical complexities of artificial articulated legs, our computer technology will be inadequate to this task for many years to come, even for very modest speeds of travel.

The historical solution for efficient land transportation has thus been wheeled vehicles on specially prepared roads or rails. The speed of transportation has been bought dearly, however, since it requires a major investment of resources to build the network of roadways and to maintain it. In the United States, we have spent approximately $175 billion (in 1978 dollars) to construct the Interstate Highway System over the last three decades, and much greater sums for state, county, and municipal highways, roads, and streets. Wheels on rails can be more efficient still, especially if all grades are moderate, permitting higher speeds in principle.

Could newer technologies permit really high speed long range rail transportation on the sur- face? Any such scheme must come to grips with four basic problems of wheels on rails: lateral stabil- ity, aerodynamic drag, intrinsic strength of the wheels, and security of the right-of-way. As train speeds become higher, irregularities in the track and motions of passengers or cargoes inside result in oscillatory motions of the railroad cars. If these motions are not to result in excessive lateral forces between the wheels and the rails or in unstable oscillations, the passenger compart- ment must be more tightly coupled to the rails and the tracks must be more precisely aligned, despite the greater lateral forces at higher speeds. This problem has restricted the actual operating speeds of the Japanese railroads to substantially less than their original design speed. The upper limit for the foreseeable future seems to be less than 200 miles per hour for advanced rail systems.

Even at speeds of 200 mph, however, trains are at a disadvantage compared to airplanes because of aerodynamic drag, since airplanes can cruise at high altitudes where lower air density reduces drag far below that felt by streamlined trains at sea level. At still higher speeds, the centrifugal stresses on wheels, in combination with the enor- mous lateral forces sustained by the wheels in holding the train on the tracks, become danger- ously close to the tensile strength of high grade steels.

These difficulties with wheels can be avoided by air cushion vehicles or by magnetic levitation. At high speeds, however, both of these technologies must still face the problem of aerodynamic drag. The necessity in air cushion vehicles for large air intake scoops and a wide base is sharply at odds with the needs of streamlining the entire vehicle. At high speeds, maintaining the stability of an air cushion vehicle to prevent accidental contact between the vehicle's structure and the roadway poses a very severe problem.

Any surface roadway or railway intended for high speeds has a perennial problem in maintaining the right-of-way secure against intrusion by pedes- trians, errant animals, or debris. Fencing is (at best) a partial solution, far from failsafe.

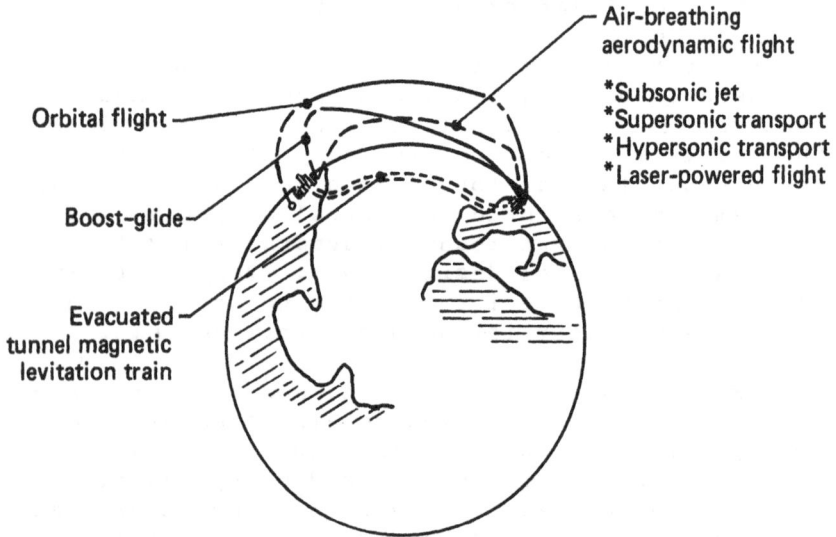

Figure 3.   Global transportation alternatives for
the turn of the century.

Pieces of metal thrown over a fence or dropped on the right-of-way from an overpass, or large tree branches blown onto the right-of-way by storms can derail trains or rupture tires, producing devastating accidents.

Transportation by waterways and by sea has been used extensively all through history. Its principal use today is for the transportation of bulk cargoes of relatively low value per unit mass or per unit volume (grains, lumber, textiles, fossil fuels). The principal advantage of aquatic and marine transportation is the fact that open water provides a level surface: little or no effort must be expended in carrying the cargo up and down the terrain. For high speeds, however, both surface and submersible vessels are severely limited by the enormous energy required to overcome hydrodynamic drag: the engines must provide the energy to displace enormous masses of water in front of the vessel to permit its passage. This difficulty can be minimized by hydrofoils, but present day experimental hydrofoils operated by the U.S. Navy can achieve only about 100 knots. Little prospect for major improvements is in sight, so hydrofoils are unlikely to present significant competition even to subsonic jets in the long-range transportation field. Ships using the air cushion effect would eliminate the hydrodynamic drag completely, but the problems of aerodynamic drag at sea level would limit marine transportation just as much as land transportation at high speeds.

It thus appears that high speed long range transportation on the surface of the earth has little future. What options remain? Besides existing subsonic and supersonic aircraft and some improvements on the existing system, five new systems could conceivably be implemented on a significant (if not fully global) scale by the turn of the century. These five technologies, shown in Figure 3, are (1) evacuated tunnel magnetic levitation trains running underground and underwater; (2) hypersonic aerodynamic flight with air-breathing engines; (3) air-breathing aerodynamic flight using laser beams for power; (4) rocket-powered flight in the atmosphere, operating in a boost-and-glide mode; and (5) rocket-powered flight into space and return. Each of these will be discussed below.

A Global Subway System

The difficulties of surface land vehicles discussed above can be overcome by going underground
to develop a novel transportation system. The
problem of security for the right-of-way and the
adversities of foul weather and large temperature
changes every day are eliminated by digging tunnels
through solid rock a few hundred feet to a few
thousand feet below the surface. Since the entire
path can be completely enclosed, the tunnel can be
evacuated to 10^{-3} of sea level pressure and density
allowing sealed vehicles to reach very high speeds
with negligible aerodynamic drag. (This pressure
and density are equivalent to those in the earth's
atmosphere at 170,000 feet above sea level.)

The problems of wheels on rails can be eliminated by substituting electromagnetic levitation
and acceleration for more conventional suspension
and propulsion methods. Each vehicle would be
equipped with several superconducting coils
carrying large currents. The currents, once
initiated, would be maintained by small cryogenic
systems. The cars of a train would then be accelerated and decelerated by magnetic fields induced
in drive coils mounted in the tunnel walls. The
timing and shape of the current pulse through each
drive coil would be controlled by microcomputers
which obtain information on the position and velocity of each car from optical and/or electronic
sensors along the tunnel. Kinetic energy pumped
into the train as it is accelerated can be re-
covered with high efficiency when the train is
decelerated by the drive coils, in a kind of
regenerative braking.

A fairly detailed conceptual design for such a
system was presented in the 1978 Macro-Engineering
Symposium.[3] The state-of-the-art in electromagnetic levitation and acceleration is perhaps most
advanced in the work of Gerard K. O'Neill, Henry
Kolm, and their associates on an electromagnetic
catapult for launching raw materials from the surface of the moon for use in space manufacturing.[4]

What sorts of speeds could such a global subway system achieve? This is principally a question of how high an acceleration passengers can

comfortably accept and of how much power can be delivered to the drive coils in the tunnel walls to accelerate the trains. In practical engineering terms, sufficient power to accelerate the trains at several g's can be provided, with higher accelerations implying higher costs for hardware. The dominant factor in the total cost of the system, however, is the cost of digging the tunnels. Thus, the question of passenger acceptance of acceleration is the real criterion. If one-third g is acceptable, then such a train system could provide a total travel time of 27.3 minutes for the 2700 statute mile trip from downtown New York to downtown Los Angeles via Dallas, reaching a peak speed at the midpoint of the journey of about 6000 mph.

A similar acceleration schedule on a tunnel following a great circle route halfway around the globe would provide semi-global travel times of 82.5 minutes, reaching a peak speed of 18,100 mph at the midpoint of the journey before commencing deceleration! At this speed, passengers would experience slightly more than one-half g centrifugal acceleration outward from the center of the earth, so their weight would be slightly less than half of normal. It has often been asserted that the fastest we could ever hope to travel on earth is orbital speed, 25,000 mph. But it is clear that tunnel trains could actually go faster than this: at 35,380 mph, passengers and cargoes would be pushed toward the top of the train at exactly one g, so that if the entire train turns upside down while accelerating through orbital speed, it could continue to accelerate to this higher speed. At that point, passengers would experience normal weight, albeit they would be seated or standing upside down! If we consider again a trip exactly halfway around the globe in such a tunnel, this peak speed could be reached with a constant acceleration of 1.27 g for 27 minutes followed by deceleration for the same time.

Because of centrifugal forces, the tunnels would have to follow great circle routes, at least in their high speed segments, and follow very precisely a smooth arc about the center of the earth. Grade changes would have to be very gradual to avoid the sensation of riding a very severe rollercoaster.

If we constrained such a system to lower accelerations, lower peak speeds, and shorter segments between stations enroute, it would still be
possible to match the travel times shown in
Figure 2 for hypersonic transports for long distances, with the major advantage that the stations
for a global subway could in almost every case be
located in the heart of the downtown regions of
cities on the subway network.

Such a system is certainly feasible from an
engineering and technological point of view by the
turn of the century. But the economic feasibility
of the system is an altogether different question.
Such a system for the United States, with major
lines connecting New York to Dallas and to Los
Angeles, major lines connecting Chicago to New
York and to Dallas, and feeder lines reaching out
to San Francisco, New Orleans, Kansas City,
Buffalo, Montreal, and Portland, Me., among other
cities, has been estimated to cost about $250
billion.

The capacity of such a system for the United
States would be five to ten times greater than the
air traffic projected for the 1990's. To make the
system economically viable, it would then appear
necessary to integrate the subway system with
other transportation and transmission systems which
can advantageously use underground tunnels. Conventional rail freight operating at high speeds
with minimal grades and no grade crossings is a
distinct possibility, especially if used for transportation of Western coal. The underground Fast
Railway System, directly connected with conventional railroads above ground, would offer a better
alternative for a coal-based electric utility industry than 100-car long unit trains running at
ten to fifteen minute intervals on the surface.

Other functions which could be integrated with
the transcontinental segments of a global subway
system include oil and gas pipelines, superconducting electrical power transmission lines, and
long distance laser communications lines. With
several such systems sharing the costs of the
initial tunneling, fares of $50 per passenger or
less for coast-to-coast travel would be possible.
But the legal problems posed by the full tangle of
anti-trust legislation and of conflicting

bureaucratic jurisdictions and regulations involved
in attempting to integrate such a diversity of
systems under a single project might easily consume decades to resolve. A further legal issue of
a new kind is the question of compensating property owners for the use of the subterranean right-
of-way, since property deeds, in common law, have
usually granted absolute rights to a land owner
to everything included in a cone down to the center
of the earth. Subways in major cities have generally been built under publicly owned land, so
little precedent is available for determining the
value of these rights.

Atmospheric Flight

In comparison with landgoing vehicles, flying
machines have always enjoyed infinite flexibility
in routing, since prepared surfaces are required
only at the departure and destination points. Sea-
going vessels, of course, enjoy this advantage too,
but the density of air being about a thousand
times less than that of water, aerodynamic drag is
much less than hydrodynamic drag, so that much
higher speeds are practical with reasonable engine
power and energy consumption rates.

Figure 4 shows the flight corridor accessible
to various types of flying machines.[5] At any given
speed, the aircraft is constrained to a certain
range of flight altitudes, either because the air
is too thin to provide sufficient lift at higher
altitudes, or because the air is so thick at lower
altitudes that friction between the air and the
aircraft skin produces excessively high temperatures. Both limits are subject to some expansion,
as some room for improvement in the design of
wings and lifting bodies still remains, and as
structural materials which can withstand still
higher temperatures become available.

The propulsion system used also depends on
airspeed and altitude of flight. At low speeds in
dense air, piston engines and propellors are
efficient and effective, although less reliable
than jet engines. As the speed of flight increases, the tips of the propellor blades
approach the speed of sound, and the shock waves
propagating backward have adverse effects on the
stability, controllability, and comfort of flight,

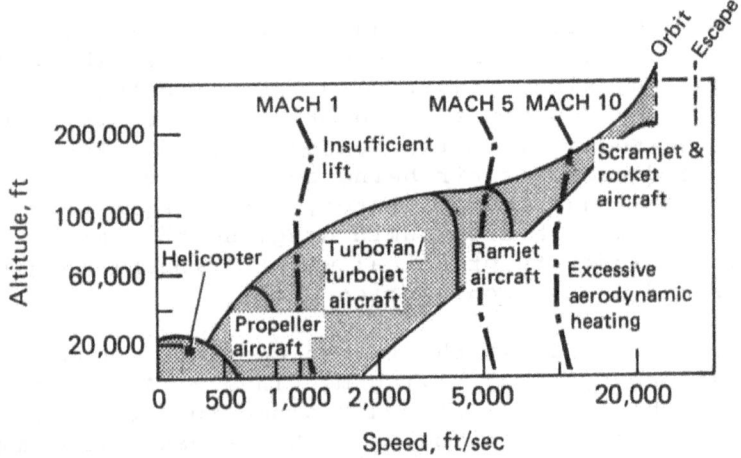

Figure 4.   Flight corridor as a function of air-
speed.   Adapted from William T. Hamilton, "Projected
Aircraft Systems Development," in The Future United
States Space Program, Vol. 38, Advances in the
Astronautical Sciences, American Astronautical
Society, San Diego, 1979.

while the overall efficiency of the propellor it-self is seriously degraded. At higher altitudes, piston engines become less and less powerful be-cause each stroke contains a smaller total mass of fuel and air and thus releases less energy per com-bustion cycle, unless the intake air is compressed by a turbocharger. At still higher altitudes, however, the temperature of the compressed air rises too high for engine components to tolerate, since the temperature depends on the compression ratio, not on the final density. If the intake air is less and less dense, the compression ratio must rise higher and higher to produce the final density needed to sustain power in the combustion cylinders.

Turbine engines perform very well over a wide range of altitudes and a significant speed range. Compression of the intake air is achieved partially by narrowing of the air flow laterally from the engine intake through the compressor into the com-bustion chambers, the very high-temperature mix-ture of combustion products expands rapidly through the exhaust nozzle to attain speeds much higher than that of the airstream past the aircraft as a whole to provide thrust. Up to airspeeds of Mach 3 or 4, then, the flow of the air/fuel mixture through the combustion chambers remains subsonic, even if the inlet and outlet speeds are supersonic. At higher speeds, however, it is impossible to keep the flow subsonic without going to very high com-pression ratios and thus to unacceptably high engine temperatures. In addition, shock waves from the supersonic flow impinging on the forward por-tions of the compressor begin to impair the efficiency (and even the mechanical integrity) of the later compressor stages.

At higher speeds, ramjets take over as the most effective engine. Having no moving parts, ramjets are far less vulnerable to internal shock waves. By narrowing down the channel for air flow through the engine, the ramjet compresses the intake air laterally and also longitudinally by slowing the flow of air through the engine. Thus, although the intake airflow may be supersonic, the flow through the ignition and combustion regions remains subsonic, again with a supersonic exhaust speed. Ignition results solely from the high temperature of the compressed air and fuel mixture.

But as the airspeed increases still further, the flow through the combustion region itself becomes supersonic, and conventional ramjets become prone to flameout. Supersonic-combustion-ramjets (usually called scramjets) must take over the task at airspeeds above Mach 6 or 7, the threshold for hypersonic flight. Note that both ramjets and scramjets cannot operate at all at speeds below Mach 2 or 3: at such low speeds, the intake air is insufficiently compressed to ignite the air/fuel mixture. Thus, aircraft designed for very high speeds must be equipped with two sets of engines: conventional turbine engines for takeoff, climb and acceleration to sufficient speed for the ramjets or scramjets to ignite. The conventional engines can then be shut down and their intakes closed over to reduce drag.

An alternative engine for high altitude, high speed flight is the rocket engine, which is intrinsically far simpler than any of those discussed above. The flow of fuel, oxidizer, mixture, and combustion products remains subsonic throughout the engine, until the exhaust gases pass through the expansion nozzle to produce thrust. The a priori liability of rocket engines in contrast to the air-breathing engines we have discussed so far is the necessity of carrying along a supply of oxidizer as well as of fuel.

Laser-Powered Flight

In view of the rising costs of fossil fuels, considerable attention has been devoted to improving the fuel efficiency of airplanes. Since the only purpose of the fuel burned in turbine engines is to heat the intake air to sufficiently high temperature to produce thrust when it re-expands through the exhaust nozzle, it is perfectly logical to eliminate the fuel altogether and to heat the compressed air by means of a remote energy source which need not be carried aboard the airplane. The idea of powering airplanes by high power lasers in space has thus been advocated by a number of studies[6] and at least one patent has been issued for this concept.[7]

Figure 5 illustrates one version of this transportation system.[8] A satellite located in a sun-synchronous orbit at 1500 km altitude at a 97°

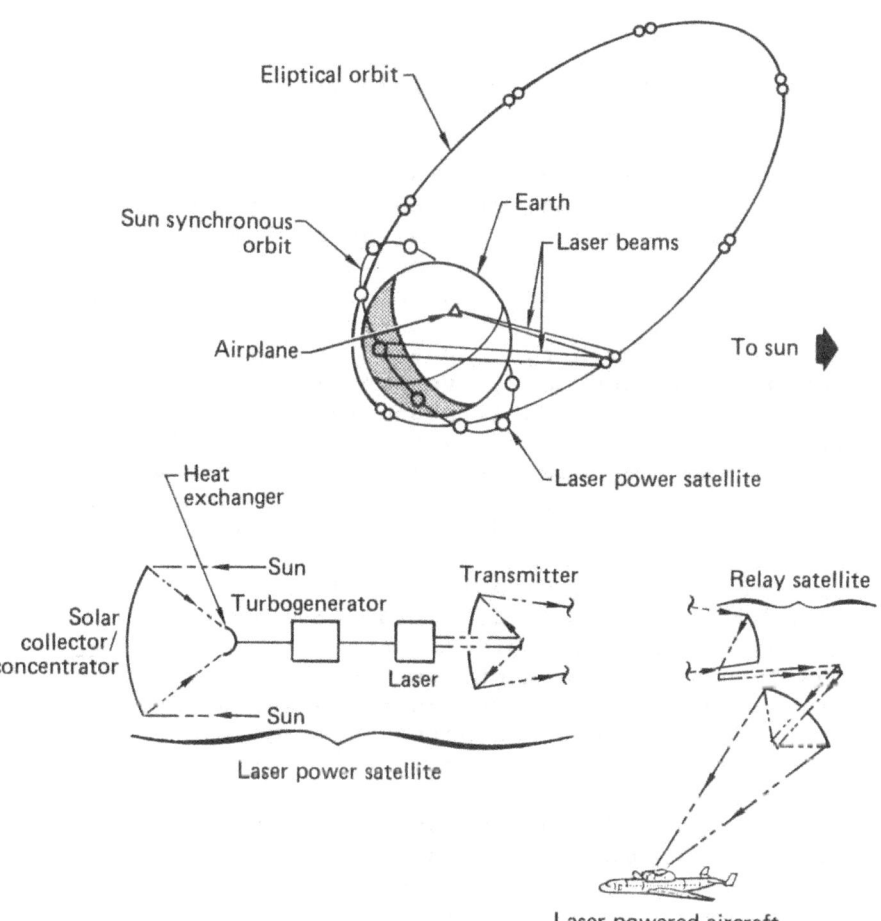

Figure 5. Schematic illustration of laser-powered
flight system. Adapted from K. Sun, A. Hertzberg,
and Wayne S. Jones, "Laser Aircraft Propulsion" in
The Future United States Space Program, Vol. 38,
Advances in the Astronautical Sciences, American
Astronautical Society, San Diego, 1979.

inclination to the equator captures solar energy, converts it into electrical energy to power a bank of high power lasers, and directs the laser beam to a relay satellite in a 4-hour elliptical orbit at a 63.4° inclination with a 500 km perigee and a 12,300 km apogee. In such an orbit, the relay satellite is above the northern hemisphere for three-fourths of each orbit, and is thus in a position to power airplanes in the part of the world where the vast majority of the air traffic presently flies.

Each airplane carries a 15-meter diameter receiver on its upper surface which focuses and directs the 10-meter diameter beam of laser radiation into a blackbody absorber cavity. The thermal energy is then transferred through a heat exchanger to the air flowing through the turbine engines, with the heat exchanger located between the compressor stage and the conventional combustor. In normal operation, the airplane would take off and climb to cruise altitude with kerosene fuel and then signal the satellite system, requesting assignment of a laser beam. Within a few seconds, a relay satellite could acquire and track the airplane and lock a laser beam onto the receiver. The kerosene flow to the engines would then be shut off allowing the airplane to cruise at stratospheric altitudes indefinitely. (Such a system would clearly have major importance for military applications, as a forward reconnaissance and observation platform, as an airborne command post, or as a deterrent strategic bomber.) When the airplane approached its destination, the flow of kerosene would be resumed and the engines reignited, the satellite system advised that the laser beam was free for reassignment, and the airplane would begin its descent. Should the laser beam be lost inadvertently due to a system failure of some kind, the fuel onboard would allow the airplane to fly 575 miles before landing, assuring a comfortable safety margin for most operations.

The fuel savings, obviously, would be substantial in comparison with conventional jet transports. Assuming a fleet of 300 laser-powered airplanes used in transcontinental and intercontinental traffic, the savings in kerosene in one year would repay the energy investment in deploying the 300 laser power satellites and 400 relay satellites needed for this system. (Each laser power satellite is estimated to have a mass of about 740 tons, while the relay satellites would be 85 tons each.) Total cost of such a system is estimated (in 1978 dollars) to be about $52 billion for the space segment and $1.6 billion for the fleet of 300 airplanes, which is slightly more than one-tenth of the size of the present domestic airline jet fleet in the United States.

Assuming a useful life of thirty years for the satellites and for the airplanes; replacement of 10% of each satellite's mass in the course of maintenance over its life; 6 to 8% annual interest burden; and equal annual payments over the life of the system, the system would be economically competitive for intercontinental flights (more than 7500 kilometers or 4680 miles) if kerosene prices rise above $1.40 per gallon, and for transcontinental flights (more than 5500 km or 3345 miles) if kerosene goes above $2.00 per gallon.

For really long range trips (up to semi-global range), the main time saving associated with laser-powered flight at subsonic speeds would be the elimination of enroute refueling stops. Sleeper service might then become very attractive on long flights. (Japan Air Lines has recently begun to offer sleeper service on some of its trans-Pacific flights).

Environmentally, laser-powered aircraft appear very attractive, since the airplanes leave no exhaust products in the stratosphere during cruise. With failsafe feedback controls on the aircraft tracking and beam pointing systems, the laser beam could be shut off within less than 100 milliseconds after wandering off the receiver on the airplane, so little or no laser radiation should ever reach the ground. In normal operation, the laser beam would sweep across the earth at the cruise speed of the airplane, so that inadvertent exposure to the beam would be very brief, and no unwitting target on the ground should ever be exposed to more than 1 Joule/cm^2 of infrared laser light.

The principal obstacle to such a system is the large initial capital investment required to deploy the space segments. The total amount of financing required is probably beyond the capability of the entire domestic airline industry, should anti-trust rules even permit them to collaborate on such a venture. Either an entirely new corporation would have to be formed to undertake the project--most likely with government guarantees or subsidies--or the federal government itself would have to carry out the program. Needless to say, the political obstacles would be substantial. Such a system would certainly be much more viable as a spinoff or subsystem of a Solar Power Satellite program based on laser transmission of power to the earth for use by the utility industry as a primary source of baseload generating capacity.

Hypersonic Flight

We saw in Figure 2 that the Concorde SST, with a cruising speed of about Mach 2.5, can provide a total flight time of about ten hours for semi-global range. To cut this time in half, it is necessary to aim at a cruising speed substantially more than twice as high because of the time spent on the ground during refueling stops and the reduced airspeeds during climb and descent. A hypersonic aircraft capable of cruising at about Mach 8, having slightly longer range than the Concorde, would reduce the time for a trip halfway around the globe to about five hours, as shown in Figure 2. While the Concorde cruises at altitudes of 40,000 to 60,000 feet above sea level, a hypersonic transport (HST) would cruise above 90,000 feet, as indicated in Figure 4.

The flight aerodynamics for a hypersonic transport are basically reasonably well understood, due in large measure to the experimental rocket airplane program, beginning with the X-1 up through the X-15. This flight regime has also been studied extensively as part of the development program for the Space Shuttle which reenters the perceptible atmosphere at speeds in excess of Mach 20, decelerating as it descends generally within the flight corridor shown in Figure 4, although during the early stages of the reentry it dips below the corridor for a brief time, requiring thick ceramic tiles for thermal protection on the undersurface which is exposed to the greatest heating.

Generally speaking, an aircraft designer will gear the design toward the lower side of the flight corridor shown in Figure 4. Although the viscous drag which produces heating is greater at the lower altitudes, the induced drag associated with the aerodynamic lift is generally less since the total surface required to generate enough lift to support a given gross weight in flight is smaller at higher air densities. If the lifting surface is smaller, the induced drag (which is generally much larger than the viscous drag) will also be smaller, and thus, the airplane will require less fuel for a given flight range. The cost of fuel economy is greater heating on the outside of the airplane, and this poses a significant but not overwhelming problem for HST design. Present thinking suggests the use of liquid hydrogen (cryogenic) as the principal fuel; because of the low temperature of the liquid hydrogen, the fuel lines from the tanks to the engines can run past the hottest parts of the aircraft skin to provide active cooling. (The fuel would also be used in the same way to cool the engine walls.) Passive techniques for keeping the cabin temperature reasonable would also be used, including multi-layered composite wall structures such as honeycombs.

The development of an operational scramjet is the principal engineering problem for hypersonic transportation. During the period 1965 to 1975, NASA conducted a Hypersonic Research Engine (HRE) Project to demonstrate the feasibility of achieving high internal performance with flight-scale scramjets. A podded, axially symmetric engine was developed and tested in wind tunnels at up to Mach 7 conditions. Due to the large nozzle expansion required at high speeds and to shock losses associated with engine struts, it was recognized that the external drag on such an engine would be very high. Thus, if an engine built in such a configuration were to be installed on a hypersonic airplane, overall performance (net thrust) would be poor. NASA had planned to test the HRE in actual flight, using the X-15 as a testbed, but the X-15 program was terminated before the HRE could be flown.

The HRE experience showed the necessity of considering engine design in conjunction with the airframe design. The presently favored concept is

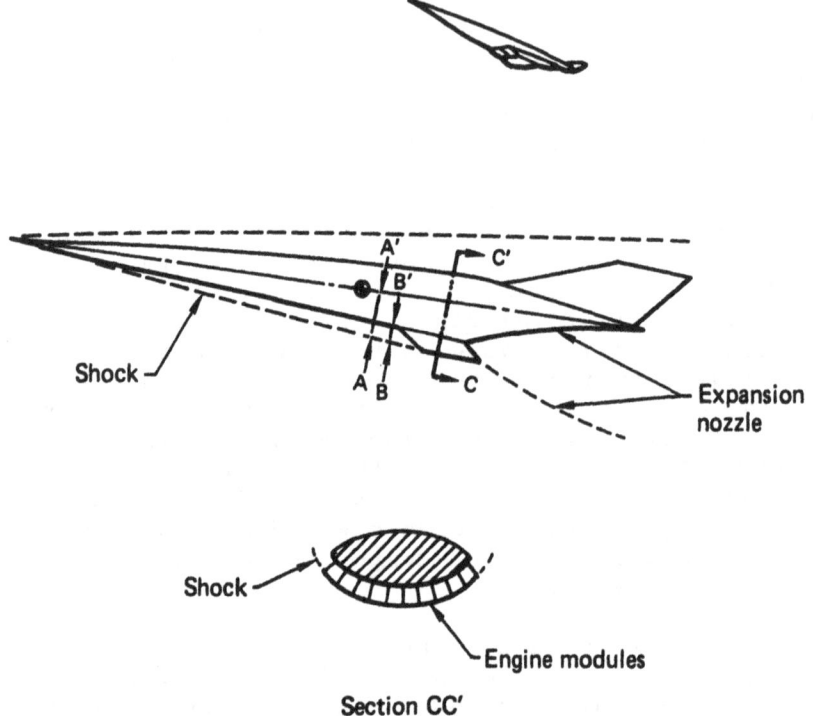

Figure 6.   Airframe-integrated scramjet concepts.
Reprinted with permission from the American Insti-
tute of Aeronautics and Astronautics, <u>Astronautics
& Aeronautics</u>, February 1978, pp. 38-48.

the "airframe-integrated scramjet," whose basic ideas are shown in Figure 6.[9] In this airplane design, the shock wave from the nose of the plane just touches the bottom of the engine inlets when the airplane is at its design cruise speed. Most of the airflow which would have passed through segment AA' has been compressed vertically into BB'. For a typical airplane dimension, the vertical height of the scramjet inlet BB' might be 1.5 to 2 feet, midline might be 9 to 12 feet, resulting in a six-fold compression of the air entering the engine. If modular engines fill up the underbody sections of an annulus projecting out from the aft portion of the fuselage to the shock surface, the engines can capture virtually all of the airflow compressed by the forward fuselage during cruise conditions. At substantially slower speeds, the engines can still scoop up a significant fraction of the compressed air flow.

The airframe afterbody is also integrated with the engines, with the aft undersurface shaped to act as the upper side of a two-dimensional expansion nozzle. According to computer calculations performed at NASA's Langley Research Center, this surface would generate some 50% of the net thrust. The shock wave trailing back from the lower engine cowling would serve, in effect, as the lower side of the expansion nozzle.

Figure 7 shows some of the internal details of a scramjet engine module. The airflow, which has already been compressed in the vertical direction by the forward section of the fuselage, is now compressed laterally by wedge-shaped sidewalls and internal struts. Fuel is injected through multiple holes just behind the narrowest point of the passage through the engine, actively cooling the sidewalls and struts (particularly if liquid hydrogen is used as the principal fuel). Because the airflow has been compressed in only one direction at a time, it should be relatively easy to compute the airflow through alternative engine designs and to verify uniformity of the air/fuel mixture by suitable positioning of the fuel injection holes.

Since the airflow has been compressed 20- or 30-fold, the temperature of the air/fuel mixture exceeds the ignition temperature, even if the flow through the combustion section is only transonic.

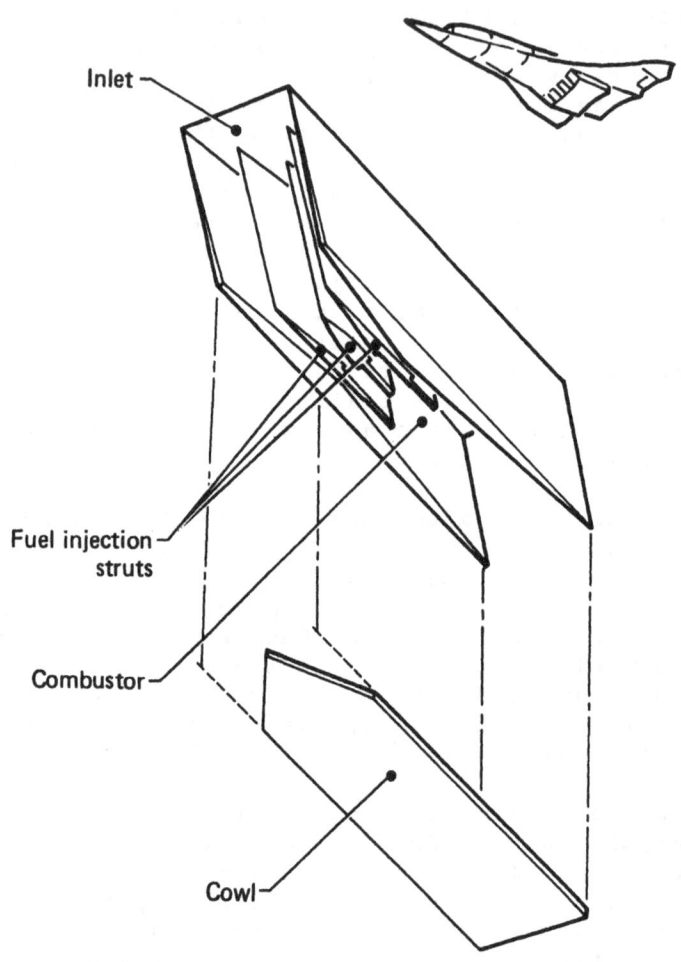

Figure 7.   Internal details of an airframe-integrated scramjet engine module, with the lower engine cowl removed.   Reprinted with permission from the American Institute of Aeronautics and Astronautics, <u>Astronautics & Aeronautics</u>, February 1978, pp. 38-48.

Thus, no glow plugs or other ignition devices should be necessary. Significant research work remains to be done, however, on the combustion processes in scramjet engines in order to assure stability of combustion and to understand how to optimize engine performance under the wide range of conditions a hypersonic transport will actually experience in routine flight operations.

Thus far, one of the best research tools available for combustion science is the shock tube. A long tube containing a uniform, prepared mixture of fuel, oxidizer, and inert gases (in whatever proportions are selected by the experimenter) is subjected to a shock wave produced by rupture of a membrane at one end of the tube. The shock wave heats the gas by rapid compression, and some time after the shock wave has passed a small element of the mixed gases, the mixture ignites. The delay between shock passage and actual ignition depends on the temperature of the mixture just behind the shock (which depends, in turn, on the strength of the shock) and on the composition of the mixture. Typically, for simple hydrocarbon fuels mixed with oxygen and argon, the ignition delay times can range from 0.1 to 100 micro-seconds as the initial post-shock temperature changes from 1000° C to 2200°C. If the airflow through the combustor in a scramjet has a speed of Mach 3, then one microsecond of travel time corresponds to one millimeter of displacement along the length of the engine. The flame front in a scramjet could thus vary in position by some 10 cm in a total engine length of 200-300 cm. Small fluctuations in inlet air densities and fuel injection rates can result in significant differences in combustion conditions and hence in combustion products and efficiency.

Small variations in the non-equilibrium conditions immediately behind the shock wave (during the first nanosecond, in some cases) may have major effects on the branches ratios for competing combustion reaction pathways for such simple fuels as methane, for which not even the equilibrium reaction rate with pure oxygen has ever been determined. During the ignition delay time, a host of chemical and pyrolytic reactions take place, producing a wide variety of intermediate species and radicals. This holds even for such a simple fuel as hydrogen in air: intermediate species include ozone, various

oxides of nitrogen, hydroxyl radicals (OH), hydro-
peroxyl radicals (HO_2), and hydrogen peroxide.
Clearly, the details of the combustion processes
will have a major effect on the engine exhaust products. Far more research needs to be done on the
details of combustion in any engine (automobile and
aircraft piston engines, turbine engines, and ramjets included), but this is especially true for the
scramjet with which we have as yet had so little
pragmatic experience.

Clearly, development of the scramjet engine to
a level comparable to the state-of-the-art for
conventional turbine engines will be a lengthy and
challenging effort. Including extensive wind
tunnel testing and in-flight tests aboard high
speed testbed vehicles, the development phase for
the scramjet engine, if adequately funded, can be
expected to last some ten to fifteen years. Major
advances in numerical simulation methods for two-
and three-dimensional hydrodynamics can be expected
in the next decade (especially using finite element
methods with moving node points), and these will
aid enormously in such a program. Once the details
of the combustion process in the engines are understood, a further difficult challenge will be to
understand the effects of the exhaust products on
the upper atmosphere, from the tropopause up to the
cruising altitudes for a hypersonic transport.

Further development work will be necessary in
materials science and in structural engineering to
provide thermal insulation and active cooling for
the aircraft interior against aerodynamic heating
during cruise, since the wing area, dry weight, and
fuel consumption can all be reduced by flying at
the lowest practical altitudes which implies highest
skin temperatures. Experience with the experimental rocket airplanes and with the Space Shuttle
development program already provide a strong basis
for flight controls at hypersonic speeds; some
modest efforts to further improve our knowledge in
this area would obviously be worthwhile in a
hypersonic transport development program.

It is clear from the controversies over the
SST, however, that a hypersonic transport would
probably be prohibited from flight at supersonic
and hypersonic speeds over land areas. Such
restrictions on the HST would imply a significant

decrease in capability from the travel times suggested in Figure 2. Some hope can be held out that such a ban might not be applied as widely against the HST because the sonic boom would be much more attenuated than that of an SST, due to the much higher cruising altitude of the HST. This question probably cannot be resolved satisfactorily before flight tests of an HST prototype. Such major uncertainties presently make the modest costs (perhaps a few billion dollars) of developing an HST highly unpalatable for the private sector. A variety of military missions for hypersonic air-planes (such as advanced reconnaissance aircraft, interceptor aircraft, strategic cruise aircraft and missiles, and highly maneuverable interceptor missiles) may well justify such a program in the public sector.

Boost-Glide Vehicles

As remarked earlier, a rocket engine is intrinsically far simpler than a scramjet or a con-ventional turbine engine, since the flow of the fuel/oxidizer mixture and of the combustion pro-ducts can be sustained at a steady level by well-designed pumps and since the flow inside the com-bustion chamber is everywhere subsonic. Thus, it would appear plausible that rocket-powered flight in the atmosphere could be developed rapidly and economically. Rocket engines could be used to boost an aircraft to a very high altitude (perhaps 100 to 200 km), after which the aircraft would glide to its destination at high speeds, using con-ventional airbreathing engines for the last 10 to 30 minutes of flight for greater ease in integra-ting such a vehicle into other air traffic approaching the destination airport. The speed of the vehicle at its peak altitude could approach orbital speed, 25,000 miles per hour, permitting semi-global travel times as short as 90 to 120 minutes.[2]

Such vehicles, however, would be subjected to very high aerodynamic heating loads during glide for periods of an hour or more, with no fuel flow available to provide active cooling, since vir-tually all propellants would be consumed during boost. Westbound flights (launched against the direction of the earth's rotation) would be sub-ject to significant reductions in payload for any

given range of travel. Development of such vehicles could indeed be possible by the early 1990's
for a cost of several billion dollars, perhaps as
much as twice the cost of a hypersonic transport
development program. As we shall see below, however, by spending slightly more money in the development phase, a much more versatile and useful
vehicle could be developed in the same time-frame,
namely an orbital transport vehicle. It thus
appears that boost-glide transports are most unlikely to be developed in the near future.

Orbital Transports

Continuing advances in rocket engine performance, in advanced structural designs, and in
higher strength materials promise major advances in
launch vehicles capable of attaining low-earth-
orbit (roughly 200 to 1000 km altitude). A number
of conceptual design studies for advanced space
shuttles show the feasibility of developing a
single-stage-to-orbit (SSTO) vehicle by the early
1990's. In contrast to the Space Shuttle, which
requires solid propellant rocket boosters and a
throwaway external tank for its cryogenic liquid
propellants, such an SSTO would be completely self-
contained, permitting rapid turnaround between
re-entry and landing and launch into orbit once
again.

During the 1980's NASA plans a wide variety
of experiments in orbit using the Space Shuttle to
develop techniques for building large structures
(up to several km across) in space; to develop new
products which can be made in the weightless
environment of orbit; and to explore the feasibility of collecting solar energy in space (where it
is available 24 hours a day, unattenuated by the
atmosphere) for transmission to the ground. Such
projects and programs, included in the phrase
"space industrialization," promise major commercial
profits in the latter part of the 1980's, with
rapid growth thereafter well into the next century.
Since the useful lifetime of each Space Shuttle
is expected to be about 100 flights, it is very
likely that all four (or five) of the Space
Shuttles presently planned will be worn out by
1990. Each Shuttle flight is expected to cost
approximately $21 million to launch a payload of up

to 32 tons, for a price per pound of payload around $325.

Present estimates for an SSTO capable of placing 250 tons into orbit suggest a launch price around $8 per pound of payload or about $4 million per launch (1978 dollars). Smaller SSTO's, more suitable for carrying large numbers of space workers and priority cargoes to and from orbit, could be expected to cost perhaps $20 per pound of payload. The R&D costs for a 10 to 12 year long program to develop such a family of vehicles would be $8 to $12 billion (1978 dollars), assuming the program starts in 1980 or 1981.[10] During the 1980's, NASA expects to launch as many as 900 Space Shuttle flights. The R&D costs would thus be amply justified by the difference in launch costs between the Shuttle ($325 per pound) and the SSTO ($20 per pound or less) during just 500 Shuttle launches.

One of the most venturesome programs of space industrialization presently under study by NASA, the Department of Energy, the aerospace companies, and numerous individuals in industry and academia is the concept of Solar Power Satellites (SPS), which were discussed in the 1978 Macro-Engineering Symposium.[11] Depending on how rapidly an SPS program were to be implemented, the total launch requirements during the mid- to late-1990's could reach several hundred thousand tons of payload per year, requiring up to 4 or 5 flights <u>daily</u> by a launch vehicle capable of delivering 250 tons into orbit. Each power satellite (with an electrical generating capacity of 10 million kw, assembled in space in 6 to 12 months, with a total mass of about 100,000 tons) would provide some $3.2 billion a year of revenues from sale of electrical power, so the economic motivations for development of the SSTO in connection with an SPS program are very strong.

Less ambitious space industrialization programs involving advanced communications, navigation, datalink services, and earth resource prospecting satellites would also justify the development of SSTO vehicles. Construction and maintenance of very large antenna arrays in geosynchronous orbit would facilitate worldwide portable personal telephone service (a Dick Tracy-style wrist telephone

	747 Airliner (HTOHL)	NASA baseline shuttle (VTOHL)	Boost-glide rocket transport[a] (VTOHL) Ref 14	Orbital rocket transports[a]			
				(VTOVL) Ref 15	(HTOHL)[b] Ref 16	(VTOHL)[c] Ref 17	
						O$_2$/RP-1/H$_2$	O$_2$/RJ-5/H$_2$
	JP-4	Solid + O$_2$/H$_2$ 2-1/2-stage	O$_2$/H$_2$	O$_2$/H$_2$ H$_2$ tanks retained	O$_2$/H$_2$ + ground accelerator	O$_2$/HC/H$_2$ 2 dual-fuel engines	
GLOW, lb	713,000	4,101,000	3,490,000	3,336,000	3,260,000	4,450,000	5,000,000
Thrust, lb	188,000	7,000,000	4,360,000	4,150,000	6,000,000 (incl. accel.)	6,050,000	6,800,000
Pr. vol., ft^3	5,200	90,000	132,000	128,000	127,000	83,000	83,000
Dry wt., lb[d]	353,000	540,000	470,000	315,000	405,000	372,000	380,000
Payload, lb[e] (or No. pass.)	103,170 / 374	52,000[e] / —	60,000 / 200	50,000[e] / 170	60,000[e] / 200	65,000[e] / 200	85,000[e] / 250

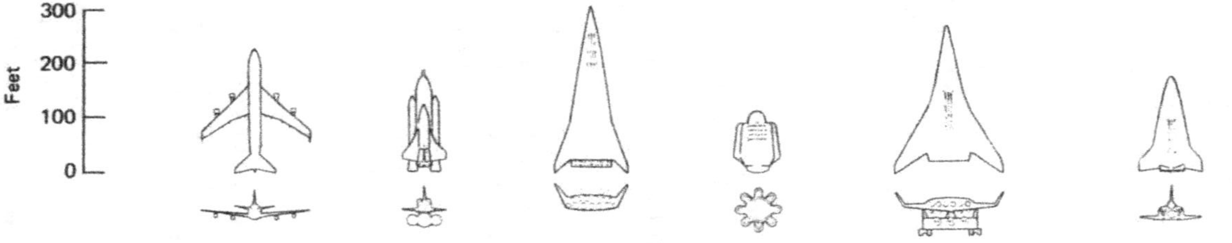

a. Designs & performance not optimized
b. Adapted & scaled from Ref 16
c. Adapted from Ref 17
d. 2-G acceleration limit
e. 100 n. mi./60^0 orbit

Figure 8. Comparison of boost-glide and orbital rocket transports to B-747 airliner and NASA Space Shuttle. Adapted from Robert Salkeld, "Global Rocket Transports: Changing Perspectives," in The Second Fifteen Years in Space, Vol. 31, Science and Technology, American Astronautical Society, San Diego, 1973.

capable of global access) and other advanced information handling capabilities.[12] The total launch
requirements for such systems may total only a few
thousand tons each year, as well as a few dozen
to a few hundred workers transported to and from
orbit each year (rather than several thousand
workers as in the case of an SPS program). Yet
revenues from such systems may easily total $20 to
$50 billion per year by the year 2000, growing to
several hundred billion dollars annually by 2010.
Given such lucrative industries in space, the
development of an inexpensive SSTO by the 1990's
is virtually inevitable.[13]

If the price per pound for launch into orbit
is $10 to $20, then the cost for a passenger ticket
into space would be between $3,000 and $10,000,
depending on crew size, load factors, utilization
rates, and degree of personal amenities provided
for each seat. But the capability of carrying a
passenger into orbit in a shuttle which can land
anywhere on earth implies a global transportation
capability. The pricetag for a trip half-way
around the world would be about the same as the
price to go into orbit, and that price is reasonable enough for some of the high-priority travel
needs I mentioned at the beginning, especially when
combined with a travel time of two hours or less
between any two points on earth.

Figure 8 compares several SSTO concepts to the
Boeing 747 airliner, to the Space Shuttle of the
1980's and to a boost-glide rocket transport. Each
vehicle shown is characterized by its takeoff/
landing operating mode, propellants used, gross
liftoff weight (GLOW), total thrust of all engines,
propellant volume, dry weight, and payload (or
passenger complement). Payloads are based on performance capabilities subject to a peak acceleration of 2 g's; for orbital vehicles, a nominal 100
nautical mile altitude (184km) orbit inclined 60°
to the equator has been assumed. (Because of the
earth's eastward rotation, launching due east from
the equator gives the best performance possible for
a given vehicle. Launching into orbits with inclinations of less than 60°, even from moderate lati-
tudes, will allow somewhat higher payloads than
those shown in Figure 8.)

While rocket engines are far simpler than turbine engines or scramjets, they are also far noisier for comparable vehicles since their thrust must be at least an order of magnitude greater. Operation of rocket engines is thus not likely to be permitted close to densely populated areas. But if the advantages of short travel times are to be realized, an SSTO should be able to operate close to urban centers, at least on one end of the trip. Horizontal landing (HL) configurations (such as the Space Shuttle) can land at conventional airports with no more noise than conventional jet transports. Such configurations are ideally suited to aerodynamic reentry and glide, with conventional subsonic turbine engines used during the last ten to thirty minutes of flight to facilitate integration with other air traffic approaching the destination airport.

Horizontal takeoff (HTO), on the other hand, poses major engineering problems for SSTO's because of the enormous mass of propellants onboard before launch. This would require massive, sturdy landing gear which would be useless in orbit and far heavier than necessary for landing, when the vehicle's tanks are empty. The high weight at liftoff would also require takeoff speeds much higher than touchdown speeds, further aggravating the landing gear problems. (Takeoff speeds of about 400 mph would be required for wings designed for a landing speed of about 150 mph.) One solution to this dilemma is shown in Figure 8. The SSTO itself is equipped with lightweight landing gear adequate for touchdown with tanks empty. For takeoff, the SSTO is placed on top of a ground accelerator, a rocket-powered dolly with heavy rolling gear which stays on the runway after bringing the SSTO to liftoff speed. Such operations, however, seem difficult to reconcile with conventional airport operations, and would require far longer than normal runways.

Another approach (not shown in Figure 8) is to take off with only partially filled tanks on lightweight landing gear, rendezvous with a very large tanker, and fill the tanks in flight. Development of a massive tanker capable of transferring cryogenic propellants at ten times the rate presently used by military aircraft is not a simple matter, and would escalate development costs for the SSTO

system significantly. As yet, however, we cannot
rule out this concept. A third approach (not
shown in Figure 8) is to carry the fully-fueled
SSTO to high altitude with a large conventional jet
aircraft and to launch the SSTO in flight.[12] Again
this approach would complicate operations and add
to the development costs of the system. Either of
these approaches, however, would permit departure
from close-in airports, with rocket ignition delayed until the SSTO is in a more remote area at
stratospheric altitudes.

Vertical takeoff/vertical landing (VTOVL) is
the operating mode familiar to fans of grade-B
science fiction movies. Rocket ships re-entering
from space come in tail first, igniting engines a
few minutes before landing, to touch down on a
large concrete blast pad amid flames and smoke.
Such vehicles, having neither landing gear nor
wings, would have somewhat better performance than
Shuttle-type vehicles. On the other hand, if
reentry is performed tail first, a more sophisticated type of rocket engine, the plug-nozzle
rocket, must be developed, to permit closing the
engine openings and to withstand the heat during
reentry.

Further developments may make some of the
above conclusions regarding HTOHL and VTOVL vehicles obsolete, since the field of SSTO concepts
is evolving rapidly and vigorously. On the basis
of published literature available at the present
time, however, it appears that a vertical takeoff/
horizontal landing (VTOHL) configuration is most
likely for the next generation of earth-to-orbit
vehicle, as shown in two alternate configurations.
in Figure 8. Instead of a pure liquid oxygen/
liquid hydrogen system, these configurations use
dual fuel engines, burning hydrocarbon fuel first,
then switching to liquid hydrogen. Liquid hydrogen is lighter than hydrocarbon fuels for the same
energy content, but requires much more volume than
hydrocarbon fuels of the same energy content.
Thus, a rocket using only hydrogen would require
much larger tanks and thus have a higher dry weight,
as well as incurring greater air resistance during
ascent. A rocket using only hydrocarbon fuel, on
the other hand, would have a greater GLOW, reducing
performance capabilities of the vehicle due to the
additional propellants needed to lift the greater

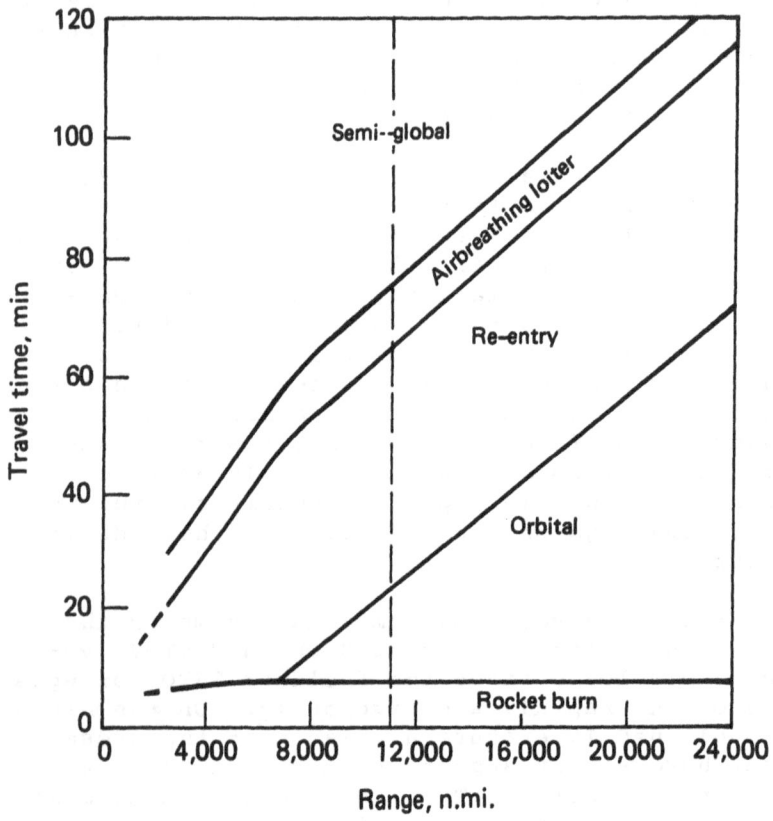

Figure 9.   Flight profile for a global rocket transport.   Adapted from Robert Salkeld, "Global Rocket Transports: Changing Perspectives," in The Second Fifteen Years in Space, Vol. 31, Science and Technology, American Astronautical Society, San Diego, 1973.

total weight initially. The optimum vehicle turns out to be a compromise between the two fuels, with the denser fuel burned off first.

While a VTOHL vehicle could not depart from a close-in airport, it can land close in and then ferry itself (using conventional turbine engines) to a more remote site for launch into space again. Figure 9 shows the flight profile for trips of various ranges. For destinations to the west of a given departure point, the SSTO would still be launched in a generally easterly direction and go around the long way. Trip ranges of up to 24,000 nautical miles are thus shown in the figure. For some pairs of cities, a second orbit may be necessary if launch performance (and payload) is not to be significantly degraded by using orbits with inclinations of more than 60°. This is most likely to happen for city pairs half-way around the globe in longitude, and on the same side of the equator. In such cases, one and a half orbits can do the trick, for a total trip time in these cases of about three hours from launch to landing.

The global transport rocket as described here would differ from a vehicle used strictly for earth-to-orbit transport primarily in the addition of conventional jet turbines and fuel tanks and the removal of extended orbital life-support systems and of orbital maneuvering rockets and propellant tanks. Little additional cost would be incurred in adapting an SSTO developed for space industrialization purposes into a global transport rocket. Depending on costs of hydrocarbon and hydrogen fuels in the 1990's (which no one can predict with high certainty), it seems that the ticket price for semi-global ranges by orbital flight would be about two to three times the cost of subsonic flights over the same range, entirely reasonable for a number of the high-priority travel needs discussed earlier. A fleet of perhaps 50 to 100 such vehicles should suffice for the antici-pated priority travel needs around the turn of the century. The construction of launch facilities for a VTOHL vehicle at sites somewhat remote from urban areas would be comparable in cost to construction of new airports. Sonic booms during launch would be a minimal problem due to the steep ascent of a VTOHL, while deceleration during reentry and glide would minimize sonic booms to those generated at very high altitudes. If necessary, approach paths can be designed to minimize the impact of even these sonic booms, since the wings of such a vehicle combined with its conventional turbine engines would permit the vehicle to veer off as much as 1100 nautical miles from the extrapolated ground track of its orbit. (This lateral maneuvering distance is called "cross range.")

Comparative Assessment

While the global subway system and the laser-powered airplane system both appear to be economically viable in the sense that anticipated revenues could amortize the initial investments, plus interest charges, the initial investments involved are very large and thus imply a great deal of risk. Such projects can be financed in a variety of ways[19], but dedicated and ardent enthusiasts would be necessary prerequisites for any of these methods of financing very large projects of such scale, making these candidates for global transportation most unlikely during the next few decades.

The boost-glide rocket transport seems to be too limited in its capacity to justify its development costs, when a small increment in the R&D costs would achieve fully orbital capability. The sonic boom throughout its flight path would pose significant restrictions on routes and thus on its economic viability. All in all, it seems that the boost-glide concept will go down in history as an amusing evolutionary dead-end.

The hypersonic transport would require a substantial R&D investment (much less than the costs of global subway or laser-powered flight, but still large compared to aircraft developed thus far) and may have difficulty proving its economic viability. Should the R&D costs be largely assumed by the military for the purpose of developing hypersonic bombers, reconnaisance planes, or interceptors, however, an HST may emerge as a spin-off.

The single-state-to-orbit vehicle, it would seem, is the leading contender for rapid global transportation around the turn of the century, since its development costs will be readily supported by the requirements of highly profitable

ventures in industrial operations in orbit beginning in the late 1980's. Use of a slightly modified SSTO for civilian global transport will be a
low-cost spin-off of this development program,
which is technologically feasible in 5 or 10 years
less time than is development of a viable scramjet.

Nonetheless, a modest research and development program in hypersonic flight would be well
worth continuing, since it would provide a concrete
focus for research in a number of vitally important
and valuable fields. Some of these include a basic
scientific understanding of the microscopic details
of combustion; advances in numerical modeling of
coupled chemical kinetics/hydrodynamics/radiation
transport; advanced aerodynamic shapes and controls;
advanced high-strength and high-temperature
materials; safe and practical use of cryogenic
fuels in aviation; interactions of exhaust products
(including water vapor) with the stratosphere and
ionosphere; propagation of sonic booms generated
at high altitudes; and advanced structural concepts
for thermal barriers.

How big can an advanced technology be and
still be beautiful? Global subways and laser-
powered flight seem too big to be viable, unless
motivated and driven by a still larger program for
other purposes. Hypersonic flight, on the other
hand, seems too small to fly--its prospective mar
kets seem too marginal to safely justify the R&D
costs, and its development timetable is too long
compared to that for an SSTO. The single-stage-to-
orbit, adapted to global transportation needs,
seems just right to work.

References

1. Kenneth Clark, <u>Civilization: A Personal View</u>, Chap. 11, "The Worship of Nature," Harper and Row, New York, 1969.

2. Robert Salkeld, "Global Rocket Transports: Changing Perspectives," in <u>The Second Fifteen Years in Space</u>, Vol. 31, Science and Technology 1973, American Astronautical Society, Tarzana, Calif. 91356.

3. Robert M. Salter, "Transplanetary Subway Systems: A Burgeoning Capability," pp.98-129, in: Frank P. Davidson, L.J. Giacoletto, and Robert Salkeld, eds., <u>Macro-Engineering and the Infrastructure of Tomorrow</u>, AAAS Selected Symposium 23, Westview Press, Boulder, CO,1978.

4. F. Chilton, B. Hibbs, H. Kolm, G.K. O'Neill, and J. Phillips, "Electromagnetic Mass Drivers," in G.K. O'Neill, ed., <u>Space Manufacturing from Non-Terrestrial Materials</u>, Vol. 37, Progress in Aeronautics and Astronautics, American Institute of Aeronautics and Astronautics, New York, 1977.

5. William T. Hamilton, "Projected Aircraft Systems Development," paper AAS 78-194, American Astronautical Society, San Diego, Calif. 1978.

6. I. Bekey, H. Mayer, and M. Wolfe, <u>Advanced Space System Concepts and Their Orbital Support Needs</u>, final report for NASA Contract NAS-W-2727, Aerospace Corp. Dec., 1976.

L.N. Myrabo, "Solar Powered Global Aerospace Transportation," AIAA Paper 78-689, IAAA/DGLR 13th International Electric Propulsion Conference, American Institute of Aeronautics and Astronautics, New York, 1978.

A. Hertzberg, and K. Sun, "Laser Aircraft Propulsion," in K. Billman, ed., <u>Third NASA Conference on Radiation Energy Conversion</u>, Vol.61, Progress in Aeronautics and Astronautics, American Astronautical Society, San Diego, CA, 1978

W.S. Jones, "Laser Powered Aircraft and Rocket Systems with Laser Energy Relay Units," in K. Billman, ed., Third NASA Conference on Radiation Energy Conversion, Vol. 61, Progress in Aeronautics and Astronautics, American Astronautical Society, San Diego, CA., 1978.

7. A.R. Kantrowitz and R.J. Rosa, "Ram Jet Powered by Laser Beam," U.S. Patent No. 3,818,700.

8. K. Sun, A. Hertzberg, and W.S. Jones, "Laser Aircraft Propulsion," Paper AAS 78-157, American Astronautical Society, San Diego, Calif., 1978.

9. Robert A. Jones and Paul W. Huber, "Toward Scramjet Aircraft," Astronautics and Aeronautics, pp. 38-48, February, 1978.

10. B.Z. Henry and C.H. Eldred, "Advanced Technology and Future Earth-to-Orbit Transportation Systems," in: Jerry Grey, ed., Space Manufacturing Facilities-II, American Institute of Aeronautics and Astronautics, New York, 1978.

11. Peter E. Glaser, "Solar Power Satellites: The Promise and Challenges," in: Frank P. Davidson, L.J. Giacoletto, and Robert Salkeld, eds., Macro-Engineering and the Infrastructure of Tomorrow, AAAS Selected Symposium 23, Westview Press, Boulder, CO., 1978.

12. I. Bekey, H. Mayer, and M. Wolfe, "Advanced Space Systems Concepts and Their Orbital Support Needs," The Aerospace Corp. Report No. ATR-76(7635)-1, NASA Contract NAS-W-2727, El Segundo, Calif., 1976.

    B.I. Edelson and W.L. Morgan, "Orbital Antenna Farms," Astronautics and Aeronautics, p. 20, September, 1977.

13. "Space Industrialization 1980-2010," Final Report, NASA Contract NAS 8-32197, Science Applications, Inc., Huntsville, AL, April 1978.

14. M.W. Hunter, II, and D.W. Fellenz, "The Hypersonic Transport: The Technology and the Potential," AIAA Paper 70-1218, American

Institute of Aeronautics and Astronautics, New York, 1970.

15. P. Bono, "Pegasus: A Design Concept for a V.I.P. Orbital/Global Rocket Transport," SAE Paper No. 64087, SAE National Aeronautics and Space Engineering Meeting, Los Angeles, CA., Oct. 7, 1964.

16. "Feasibility Study of Single-Stage-to-Orbit Vehicle," Final Report, The Boeing Company, Air Force Contract F04701-72-C-0418, Seattle, WA, January, 1973.

17. Robert Salkeld, "Mixed-Mode Propulsion for the Space Shuttle," Astronautics and Aeronautics, pp. 52-58, August 1971.

    Robert Salkeld and R. Beichel, "Reusable One-Stage-to-Orbit Shuttles: Brightening Prospects" Astronautics and Aeronautics, June, 1973.

18. Robert Salkeld and Robert S. Skulsky, "Air Launch for Space Shuttles," Acta Astronautica 2 pp. 703-713 (1975).

Solar Power Satellites and Automation in Space

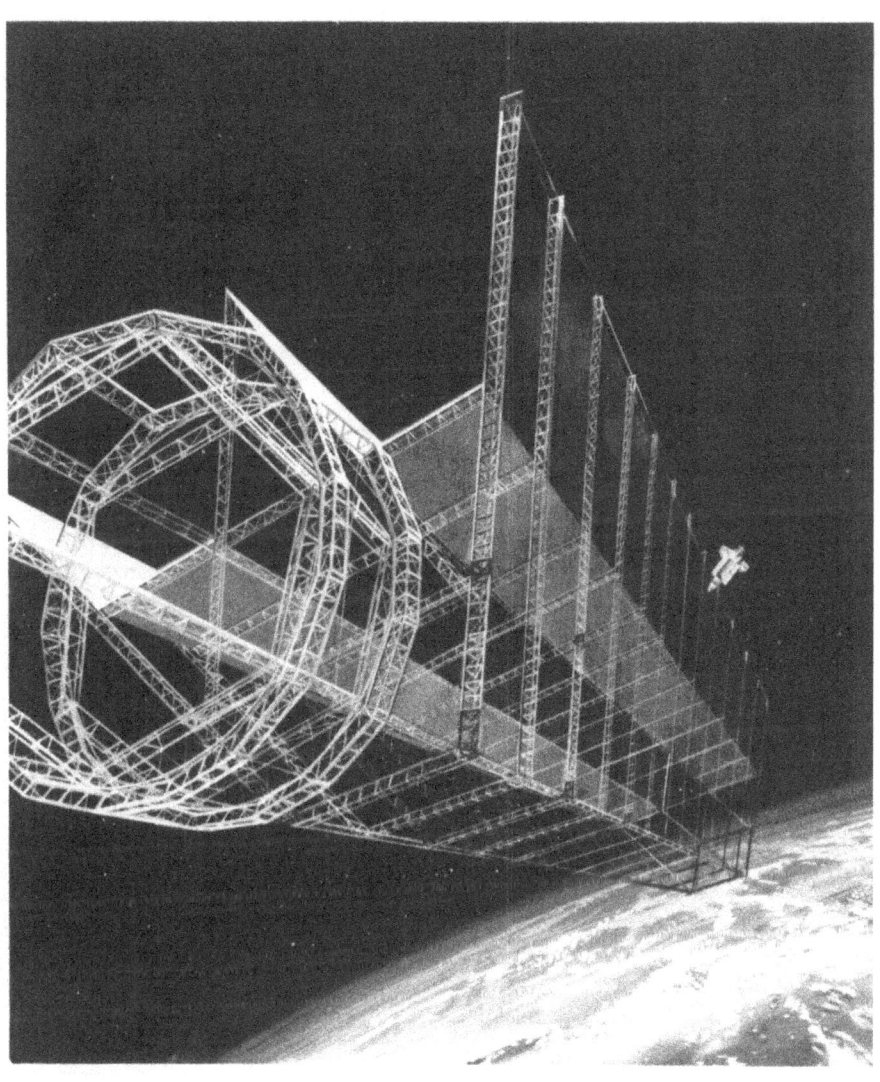

Space Shuttle can deliver both the materials and
the machinery required to build large space structures, such as this demonstration satellite solar
power station. After being fabricated and assembled
in low earth orbit, a newer station would be
transferred to its permanent place in geosynchronous
orbit (about 22,000 miles out in space) where it
would beam a continuous stream of microwave energy
to earth receivers, which would convert the energy
to electricity. Courtesy of Rockwell International,
Space Division.

Overview

SOLARES is a space-terrestrial solar energy system consisting of a set of orbiting mirrors that would provide nearly continuous reflected sunlight to a world-distributed set of solar conversion sites. Kenneth W. Billman, William P. Gilbreath, and Stuart W. Bowen argue that this system promises renewable solar energy which will be economically competitive with alternative fossil and nuclear sources. Related studies are cited to support the technical and economic feasibility of the SOLARES system. Questions of environmental impact and social-political acceptance remain to be answered.

_Kenneth W. Billman,
William P. Gilbreath and Stuart W. Bowen_

15. Solar Energy Economics: Orbiting Reflectors for World Energy

Introduction

The natural storage processes for solar energy have, over the eons, produced an inexpensive supply of fossil fuels which has prevented the economic acceptability of direct solar energy use. Today, when these reserves of stored solar energy are perceived to be inadequate for future generations, when their increased cost ferments political and economic instability, and finally, when the quality of life can be degraded by their mining, transport and burning, the search for economic methods of using the "free" and perpetually available solar energy source has intensified.

This paper outlines the recent study we have made on a space-terrestrial solar energy system, which we call SOLARES, consisting of a set of orbiting mirrors that provide nearly continuous reflected sunlight to a world-distributed set of solar conversion sites. As will be shown, this added insolation dramatically reduces the cost of direct solar-electric production. Additionally, the magnitude of energy supply on a global scale can be substantial. This combination of significant and renewable energy supply to the world community, economically competitive to alternative fossil and nuclear sources, is the promise of the SOLARES concept.

Accompanying this promise, there are, of course, many challenges which must be met. Some have been identified in our technical assessment. Others will certainly emerge if SOLARES becomes an accepted NASA program and in-depth engineering,

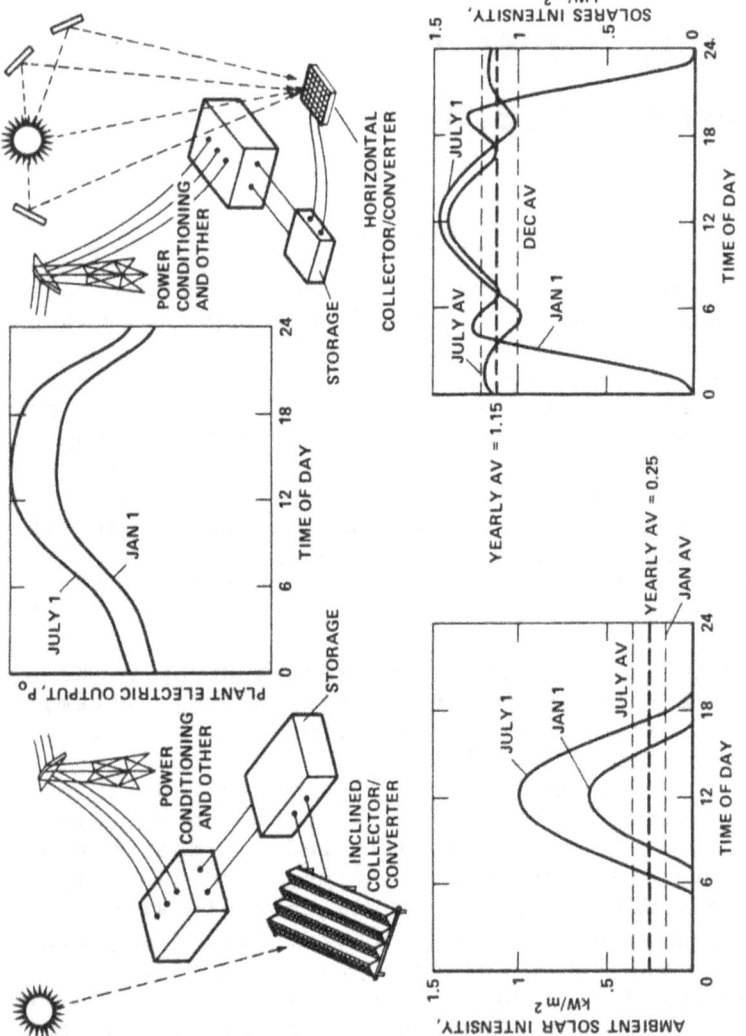

Figure 1. An illustration showing the decreased collector/converter, storage and other costs of a solar farm resulting from orbiting mirror insolation.

systems, and economic analyses are performed.

Without these in hand, it may appear presumptuous to accord SOLARES the promise of significant economic improvement to solar-electric production. However, as will be seen, the predominant cost element of the system is the solar terrestrial converters. Fortunately, this element has been intensively studied by others, for ambient sunlight use, and firm engineering and circa 1990 costs have been established. The secondary SOLARES cost element is presently believed to be the space system--resulting from the simplicity, efficiency, and especially the low mass density of the passive solar reflectors which we, and others, envision as suitable. It is in this space element where one could anticipate sizable change with deeper analysis. But, as will be shown, even five-to-ten-fold increases in its cost will still allow SOLARES to be competitive to alternative coal, oil, and nuclear electric power sources! It is this insensitivity to space costs, coupled with the firmer terrestrial costs -- which, by scientific analysis, can be demonstrated to be significantly reduced with the SOLARES approach -- that lends credence to our assessment of the importance of this new energy alternative.

The impact of SOLARES on the economics of solar-electric generation can be seen in Figure 1. We consider here two solar farms: one operating with normal sunlight, the other with normal sunlight augmented with space mirror-reflected insolation. If these systems are to compete with alternatives, they must produce baseload (continuous) power. Typical power output or "demand" curves are shown for 24-hour periods for summer and winter. To meet these, each system must provide expensive energy storage. From the normal solar insolation curve (lower left) it can be seen that this is substantial: demanding about 14 hours of capacity. It is for this reason that economics normally preclude a baseload solar energy system. Careful analysis of the SOLARES insolation curves show that although some eclipsing of the mirrors by the earth occurs around local midnight at the site, the required storage is reduced typically by a factor of five. Secondly, the "diluteness" or low average energy density of ambient sunlight is noted. The yearly average for a good solar farm site is only about 0.25 kW/m^2, with a two-fold seasonal variation.

This low density demands an extremely large area of
expensive solar collectors and converters for significant energy output. The use of concentrators
to reduce the area of expensive converters required
is unfortunately accompanied by the need for
tracking the sun which results in an even more
costly and complicated ground site. In contrast,
the SOLARES mirror system is sized to provide a
yearly average 1.14 kW/m^2 (typical high-noon desert
intensity), with a significantly lower seasonal
variation of about $\pm$ 10%. Hence, to produce the
same electric output, the SOLARES solar conversion
array is reduced by a factor of 1.14/0.25$\simeq$4.6.

This potential reduction of nearly 5 in each
of the most costly items of solar farming--storage
and solar converters--constitutes the promise of
SOLARES. The challenge, in the main, is to accomplish this by means of a space system whose cost
does not erode this economic advantage. It is our
belief that our design of the SOLARES mirror system
which makes extensive use of the thin film, large
area structures which appear possible in the
weightlessness of space, the use of solar sailing
to reduce space transport costs and other innova-
tions should meet this challenge.

This report will address the SOLARES concept
under those four qualities which must be possessed
by any new energy alternative in order to achieve
acceptance: TECHNICAL FEASIBILITY, SIGNIFICANT AND
RENEWABLE ENERGY IMPACT, ECONOMIC FEASIBILITY, AND
SOCIAL/POLITICAL ACCEPTABILITY.

Technical Feasibility - A SOLARES Challenge

The initial technical assessment of the
SOLARES concept[1], and subsequent detailed study[2-3]
have evolved a space-terrestrial system which, in
principle, will work. As will be seen, however,
its requisite magnitude -- to make a significant
impact on world energy needs -- is large and has
few parallels in history. Fortunately, the system
consists of highly redundant space and terrestrial
elements, amenable to the economics of mass production and most importantly, capable of early and
decisive proof-of-principal testing with the Space
Shuttle and in normal solar conversion simulations.
We review here the technical aspects of these space
and terrestrial components.

The Space System

An artist's sketch of the orbiting mirror system as seen from space, Figure 2, illustrates the method of providing nearly continuous, enhanced solar energy to a typical ground site. Each mirror orbiting the earth with a particular period, is continually oriented when over a site to reflect its solar image onto a common ground spot. The image size is prescribed by geometric optics, having a minimal circular diameter of $D_M + h\alpha$, where D_M is the mirror diameter, h its altitude above the earth's surface, and α is the subtense angle of the sun (0.0093 rad $\simeq$ 1/100 rad). For a non-zenith mirror, this image is increased in size and is elliptical in shape. The superposition of such intensity profiles from all the mirrors over the site constitutes the SOLARES mirror contribution to the conversion farm; additionally, during the daylight hours, the ambient or direct sunlight is used. Detailed numerical modeling has shown that the terrestrial intensity profile resulting from all mirrors above 30° from the horizon, the "useful viewing circle" or UVC, is relatively uniform over a circle of diameter $h\alpha \simeq h/100$ and diminishes rapidly beyond this. This "$h\alpha$-circle" thus sets the scale of a SOLARES conversion farm and, interestingly, also sets the scale for the total mirror area needed in the UVC to provide a given value of average radiant energy density delivered to the site. If the latter is chosen as 1 kW/m^2 the oversite mirror area is found roughly equal to 3 times the $h\alpha$-circle area. This leads to certain important conclusions. First, the use of a single, geostationary mirror which would remain fixed in position above a single site appears excessive in land and mirror areas required since the orbit altitude of 35,800 km would give a spot size of about 333 km or the size of the state of Maine! Lower orbits ranging in value from the lower limit of about 1,000 km (period $\simeq$ 2 hr), dictated by atmospheric drag on the low ballistic coefficient mirrors, to an upper limit of about 6,500 km (period $\simeq$ 4 hr), dictated by the predicted world energy needs circa 1990, are of more significance. The second resulting conclusion is illustrated in Figure 3, viz. a world-wide orbital mirror set is necessary so that as some mirrors leave the UVC of a given site, others enter it and hence an uninterrupted energy supply is maintained. The added

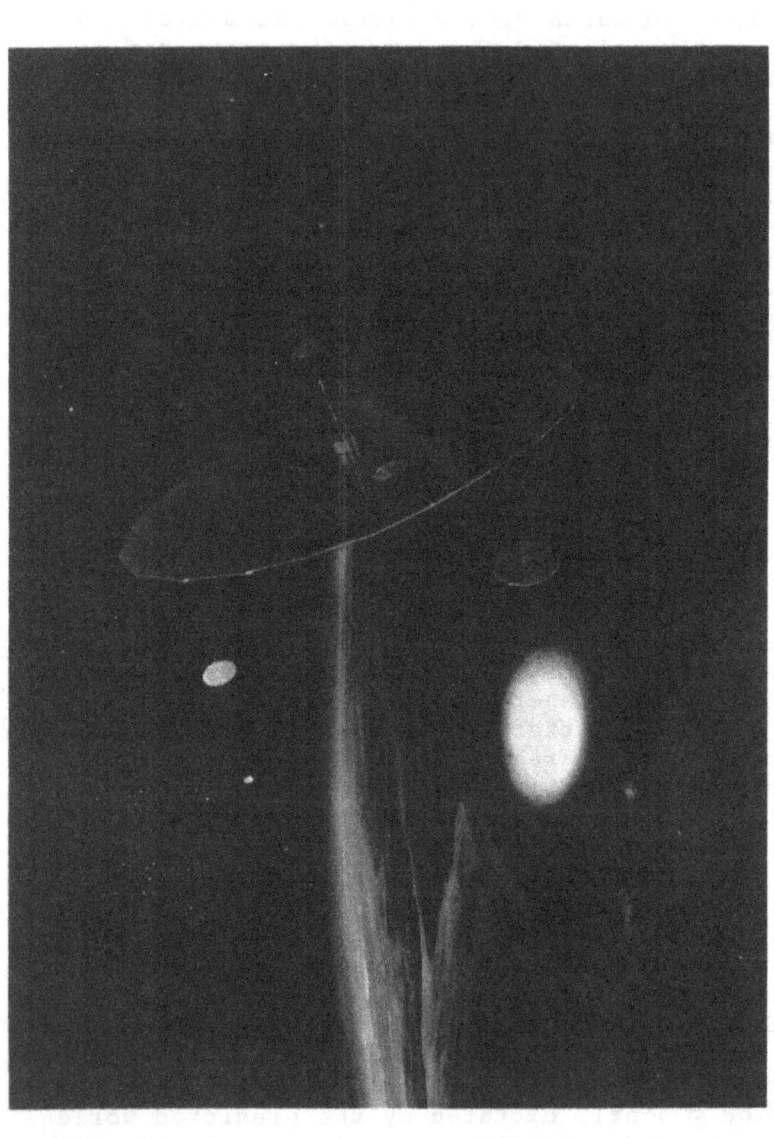

Figure 2. Artist's concept of a space view of the orbiting mirror system.

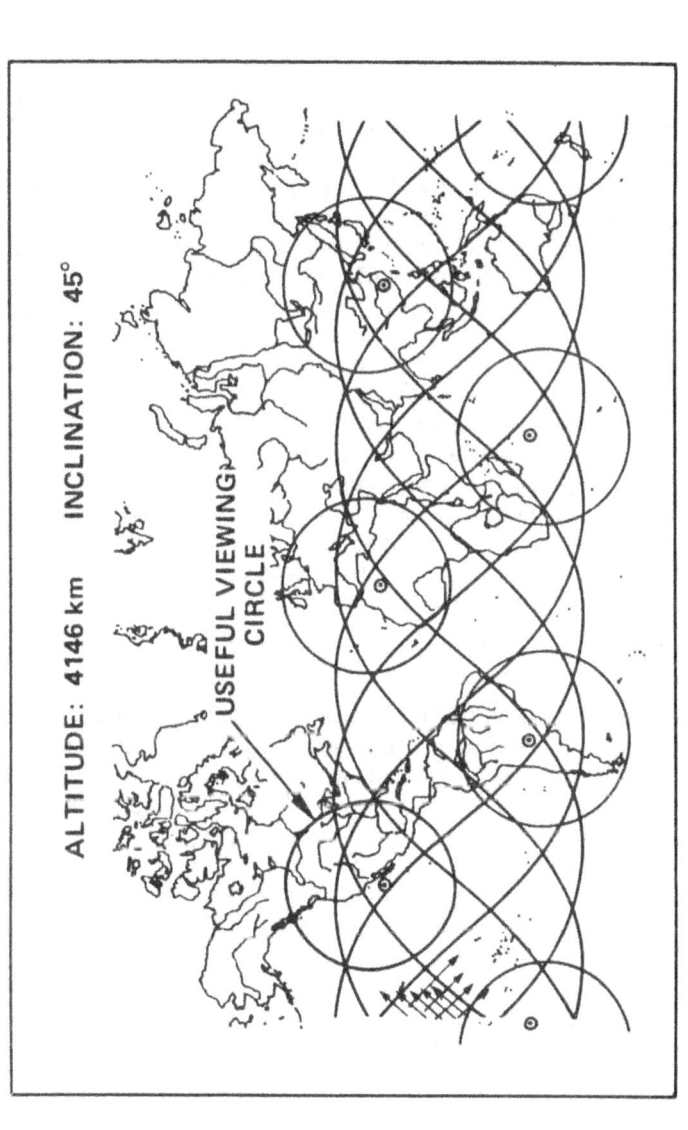

Figure 3. Ground tracks of a typical mirror showing its daily coverage of the Earth. Other mirrors follow similar, interleaved orbits.

complexity of such a multiple mirror set over that
of a single mirror is believed to be minimal since
all such mirrors are "free-flyers", that is, independent and individually controlled in either case.
Additionally, many sites may now be serviced simultaneously. The maximum number is fixed by choice
of UVC size, depending on altitude and the chosen
acceptance angle, and the allowable reorientation
torques which may be applied to the low-mass mirror
structure. These are necessary, of course, to prepare a mirror for service to a site during the time
it spends orbiting between it and the prior site.
This allowable site number ranges between 16, for
low altitude mirrors (1196 km) serving equatorial
sites, and 4, for high altitudes (6384 km); the
former would produce electric power at a level of
about 13 GWe/site while the latter, larger sites
would deliver about 375 GWe each. For reference,
current usage in the United States is about 250
GWe. It is interesting to note that the SOLARES
system can be expanded, to meet increasing world
energy demand, by raising the altitude of the
mirrors and simultaneously increasing the conversion site radius to accommodate the resulting larger hα-circle. Such a growth scenario will be
explored later in this report.

Orbits

Many special orbit types, such as geostationary, polar and sun-synchronous, have been considered; each possesses certain advantages and disadvantages. Of most utility are circular orbits of
inclination somewhat larger than that of the latitude of the northernmost (or southernmost) sites
to be serviced. We exclusively consider here two
particular cases: those at about 40° inclination,
servicing sites at 30° latitude and those at 0°
servicing equatorial sites. The latter, equatorial orbits, are of particular interest since they
make most effective use of the mirrors -- each
mirror passes over each equatorial site on each
successive orbit. Hence, these require a smaller
mirror set (or space expense) to deliver the prescribed insolation.

Insolation

To quantitatively determine the mirror system
reflective area required, extensive numerical

computation has been made. The physical model used[3] places mirrors of unit area into prescribed orbital positions over the site and sums the transmitted intensities of each to the prescribed (latitude and longitude) site for each incremental time step during the 24 hours of the particular date being examined. All geometric factors, the equation of time, seasonal variation of the effective solar constant, eclipsing (earth shadowing) of certain mirrors, and the absorption and scattering by the atmosphere are rigorously included. Typical outputs from these computations are illustrated in Figure 4 for the case of a 30° latitude site and mirrors at 6384 km altitude with 40° inclination. On the left is seen the spatial radiant intensity profile, showing the uniformly illuminated $h\alpha$-circle with rather sharp decrease in intensity outside it. On the right, two temporal plots illustrate the variation of the value of this central $h\alpha$-circle intensity during typical summer and winter days. Ambient sunlight is also shown. Interestingly, it is seen that although the insolation provided by the mirrors decreases around noon, since the effective mirror area is geometrically lessened at this time, the ambient solar insolation is large and hence the total insolation available to a horizontal ("flat-plate") conversion system is reasonably constant. The effect of eclipsing during the winter months is seen occurring around local midnight in the lower curve. As discussed earlier, minimal storage (cross-hatched area above the dashed demand curve) suffices to provide continuous baseload output during these periods (cross-hatched area under the demand curve).

From these insolation curves, together with the determination of the fraction of the space system mirror area which is above the site, an accurate assessment of the total space system mirror areas required to give a yearly averaged site radiant intensity of 1.14 kW/m^2 is determined. As seen in Table 1 for various inclined and equatorial orbits, the total areas are large, ranging from about 6,000 km^2 to 70,000 km^2. If 1 km^2 mirrors were used, this would dictate, of course, a similar number of mirror satellites. Indeed, if, as will be shown, this system can nearly duplicate in output that of the existing world system of hydro-, fossil and nuclear-electric power plants, it is not surprising that a large system is

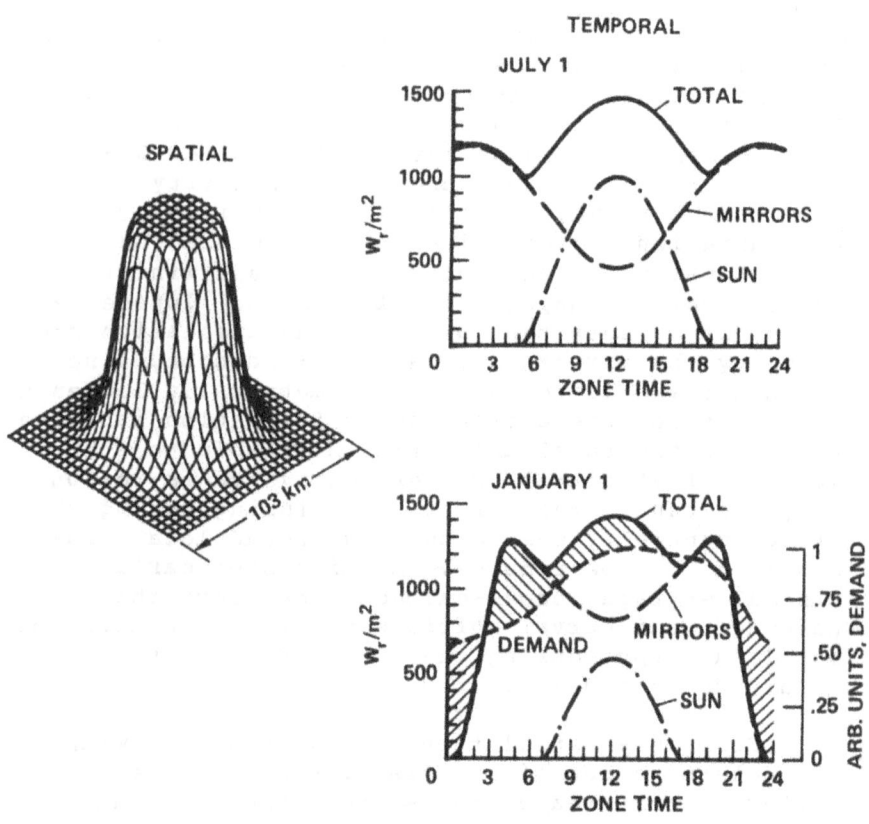

Figure 4.   Spatial insolation profile (left) and intensities at site center for typical summer and winter days.

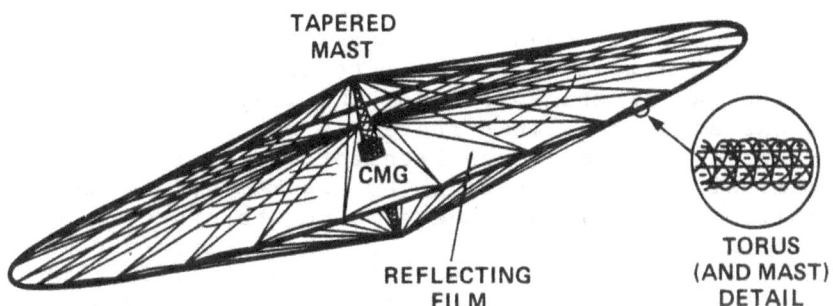

Figure 5. Illustration of the Ames SOLARES mirror configuration.

Table 1. Total space system mirror areas required to provide a nominal 1.14kW/m² on sites[a].

Mirror altitude, 10^3 km	Orbit inclination, deg	Orbit period, hr	Total mirror area, 10^3 km²	$h\alpha$-circle diameter, km
1.18	0 (Equatorial)	1.82	6.4	10.95
2.10	0 (Equatorial)	2.16	10.4	19.52
4.14	0 (Equatorial)	2.98	21.6	38.5
6.38	0 (Equatorial)	3.98	36.4	59.34
1.20	45	1.82	29.1[b]	11.12
2.11	40	2.16	25.8	19.61
4.14	40	2.98	45.8	38.56
6.38	40	3.98	69.0	59.38

[a]Includes ambient insolation.
[b]Eclipsing raises this area above that for the 2110 km orbit.

necessary. The important question is not scale,
but rather those of technical feasibility and economic desirability of achieving it. Both, we
believe, can be answered affirmatively.

Mirror Design

Of critical importance to minimization of the
space cost component of SOLARES is to reduce the
mass which must be placed into orbit. Since the
mirror areas are specified, the areal density of
these structures must therefore be minimized. Fortunately, the space environment is favorable to
this since an orbiting object is weightless and, at
sufficient altitude, there is no structure loading
by atmospheric viscous drag. These factors point
to the possibility of very large area $(1-50 \text{ km}^2)$
low average mass density $(10-200 \text{ gm/m}^2)$ reflectors
which can maintain their required optical surface
flatness under the residual forces: radiation
pressure, gravity gradients, and the inertial
loading due to intra-and inter-site orientational
torques.

Our "Mark I" design of such a structure is
schematically shown in Figure 5. It consists of a
thin (ca. 2 μm) polymeric film, overcoated
(ca. 0.1 μm) on each side with highly reflective
aluminum, tautly supported by a bicycle-like support structure consisting of an outer toroidal rim,
a tapered mast (or hub) and guy wires (spokes)
between these. This graphite-epoxy, thin-walled
structure is dimensionally sized at each potential
mirror operational altitude, that is, for the calculated residual forces, to maintain the specified
mirror flatness (in angular measure equal to about
$\alpha/10$ where α is the subtense angle of the sun)
or, equivalently, the film tension. The torus
cross-section and wall thickness is iteratively
determined to withstand local buckling and in-
plane buckling under the combined mirror tension
and torques. The mast wall thickness is determined
to withstand local buckling primarily due to the
imposed torques. An optimal mast length has been
found to be approximately $a/5$, where a is the
mirror radius.

At the expense of added constructural complexity, but with resultant decrease in overall mass
density, the reflecting film may be divided into

smaller panels which are tensioned onto a supportive, in-plane grid which is maintained planar by
means of additional guy wires fastened to the mast.
At each operational altitude, a panel size can be
determined which minimizes the overall mass density
σ of this "optimal panel" mirror.

Finally, the means for providing orientational
torques has been determined by study of oversite
acceleration requirement simulations for a large
number of representative mirror passes. These are
found to be antisymmetric in time, that is, positive acceleration as the mirror moves from one UVC
horizon to the site zenith and a corresponding but
negative acceleration as it moves from zenith to
the exiting horizon. Thus, the average torques are
nearly zero and much smaller than the required peak
torques. This characteristic allows advantageous
use of reaction wheels, combined with energy storage, which obviates the need for thrusters and
their undesirable need for fuel resupply. The
masses of these reaction and storage wheel systems,
centrally located on the mirror structure, together
with a small photovoltaic power supply, and on-
board control electronics are added to those of the
reflecting film and support structure to evaluate
the total mirror mass and the average mass per unit
mirror area.

As seen in Figure 6, the areal mass densities
can be remarkably small. For smaller mirrors, it
approaches that of the thin film reflector itself
(ca. 3.5 gm/m^2) for modest altitudes. However, at
a given altitude, the penalty for increasing the
mirror radius, and therefore, area, is apparent.
It is also seen that because of increased mirror
loading at lower altitudes, the mirrors must be
more robust and thus of higher density.

For a nominal mirror reference in this report
(baseline value), it is seen that the density of a
single panel mirror, operating at 4146 km, of
radius a = 500m and area $\pi/4$km^2, is σ_b=8.76 gm/m^2
or about one-fifth that of newsprint. It is also
found that at this altitude, optimum panel mirrors
of approximate areas of 30-, 40-, and 50-km^2 have
respective values of σ= 10 σ_b, 15 σ_b, and 20 σ_b.
These large area mirrors, which would reduce the
number of SOLARES satellites to as few as 916 for
4146 km, will be considered later.

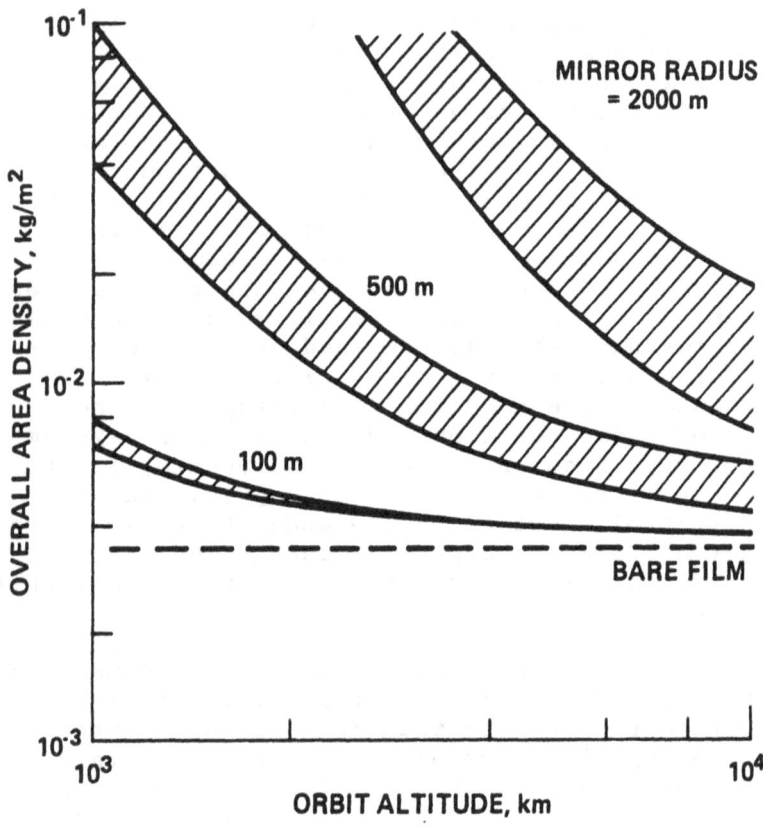

Figure 6.   Variation of the Ames mirror overall area density with service altitude and mirror radius as a parameter.

Finally, it should be noted that this mirror design, and subsequent conclusions derived from it, are subject to considerable refinement and improvement resulting from studies which are just beginning in this new class of thin film space structures. One such study,[4] by John Hedgepeth of ASTRO Research Corporation, has recently evolved a point design for a similar single panel 1 km structure, exclusive of orientational controls, at substantially lower areal density than that called for by our design. Significantly, it also includes means for thermal compensation and for mirror packaging. It is believed that such a structure could be launched in a single space shuttle and deployed, rather than constructed, in space. This would allow significant and early testing on the most uncertain element of SOLARES - the space component.

Space Transport and Manufacturing

Assessment of the required mirror area, together with our current model mirror densities, allows the determination of the total mirror system mass as a function of operational altitude. As seen in Figure 7, the system mass exhibits a broad minimum in the orbit range of 2,000-4,000 km. Although these masses are large, it is notable that the important measure of a space energy system-- mass per unit of electric power produced on the ground -- will be found to be astonishingly low, typically 0.2-0.3 kg/kW for 4146 km equatorial and inclined orbits, respectively.

The space transportation system requirements are basically determined by the mass of the mirror system to be placed in orbit. A number of studies have shown that to install systems in the mass range needed for SOLARES, it is an economic necessity to use a new generation of launch vehicles of the heavy lift class (HLLV). Such completely reusable "space freighters" are predicted to be capable of carrying 400 metric-ton payloads to low earth orbit for as little as $10/kg in the 1990's, in contrast to the expected space shuttle charge of about $1,000/kg.

Since the launch costs are both mass and rate dependent, the most difficult task in assessing the ultimate cost of the SOLARES space system, or ultimately the cost per mirror, has been in this

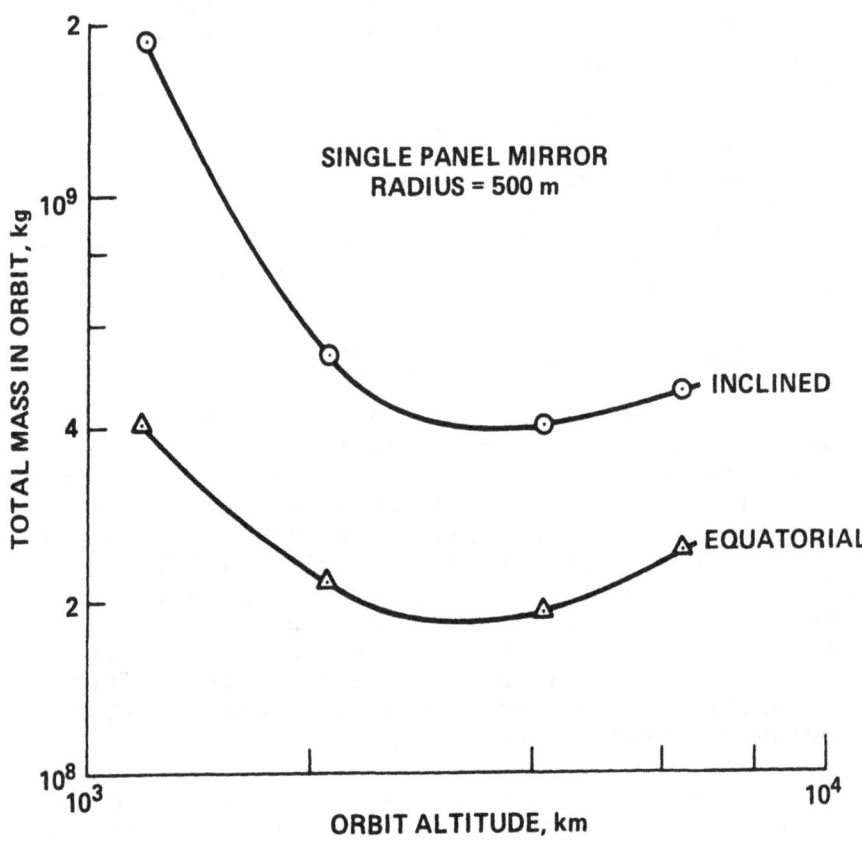

Figure 7. Total mass of the mirror system as a
function of service altitude for the inclined and
equatorial orbits.

area. Fortunately, extensive studies have been
made[5] on a system, the Satellite Power System (SPS)
with similar requirements. Our analysis of their
HLLV shows that 370 metric-ton payloads could be
delivered to 800 km altitude. For a 3×10^8 kg
SOLARES system, this would require approximately
811 launches, or about 150 launches per year for a
6-year implementation period. Because this is
lower than the 365 launches/year estimated for the
SPS, we have accordingly assessed a higher launch
cost of $59/kg to SOLARES in contrast to the SPS
value of $39/kg.

As shown in Figure 8, three other primary
transport needs will be required: (1) a personnel
launch vehicle (PLV) to move construction, support,
and maintenance crews to and from low earth orbit
(LEO), (2) solar sailing from LEO to mirror operational orbit and (3) an inter-orbital transfer vehicle (ITV) to move personnel from LEO to the
mirror operational orbits for maintenance and repair operations. The PLV would likely be a deriva-
tive of the Space Shuttle, sized to transport a
crew of 75. The figure bases construction crew
needs on the assumption that the mirrors are largely space manufactured. If it is possible to simply
deploy the mirrors, then space crew requirements
and the number of PLV's would be greatly reduced.

A key element to space cost reduction appears
to be the possible use of solar radiation pressure
to provide transport ("solar sailing"[6]) of the
mirrors from LEO to their operational orbit. Although this may be faulted in that it is a slow
process, typically requiring about 65 days to
reach 4,000 km, this is a small "offset" in a
system requiring 6 years to construct. Most importantly, it avoids the development of an inter-
orbit cargo transport system and the additional
orbital burden of it plus its requisite fuel which
would have to be carried by the HLLV. Note that
the mirrors will be placed into a collision-avoid-
ance configuration, viz., spaced within a given
circular orbit and all such orbits will be of
slightly different radius. This will avoid the
possibility of mirror collision at the "cross-over"
points which were indicated in Figure 3. It should
also be noted that the solar sailing option should
allow a convenient means to produce system expansion by mirror orbit raising, as discussed earlier.

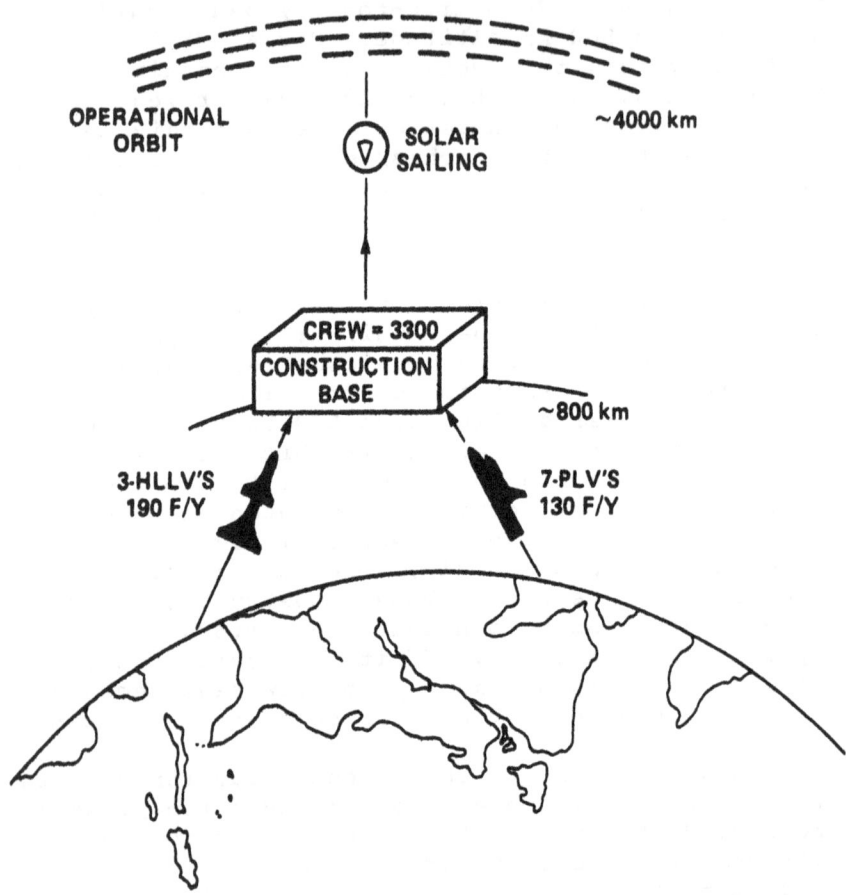

Figure 8. Schematic illustrating the three major transportation elements: HLLV, PLV, and solar sailing and the low Earth orbit construction base which may be required for in-space mirror fabrication.

The Terrestrial Conversion System

A unique property of the SOLARES system is its
potential to deliver energy in a form suitable to
many direct uses and with the economic advantage
associated with continuity and high average energy
density, as discussed earlier. Furthermore, as
illustrated in Figure 9, it can do this simultaneously, namely, those mirrors over industrial
nations can serve their needs, such as electric and
fuel production, while at the same time those
mirrors over agrarian nations could satisfy such
needs as desalination and pumping of irrigation
water, fertilizer production, etc. Additionally,
in times of emergency, a small fraction of the
mirrors could be used to provide large area illumination, frost prevention or snow cover removal.

Although these many uses are potentially important and exciting, limits on the present study
have not allowed a true assessment of the cost/
benefits of this general SOLARES application. It
is clear, as illustrated in Figure 10, that a
SOLARES conversion farm will consist of a "mix" of
solar conversion technologies which optimize the
economics of this most costly element of SOLARES
while minimizing the negative environmental effects.
Within the hα-circle, the highest unit area cost
converters, such as solar cells or for special
orbits, solar concentrators and converters, will be
located. An extensive array of solar-intense
biomass growth facilities, desalinators, fuel
production facilities, and industries requiring
process heat will undoubtedly be placed in the
lower intensity outer region of illumination. With
controlled sensible heat release, windmills operating with the high efficiency possible with uni-
directional and uni-velocity winds, will provide
economic means to pump fluids. Finally, the output electricity, fuel, heat, biomass and other
products will be linked to the users by a combination of transportation means, such as the super-
tankers shown.

It is thus clear that the SOLARES site will be
a testing, and hopefully, proving ground for the
many developments in solar conversion technology.
Research dollars spent on this element of SOLARES
will notably benefit not only centralized power
needs, so necessary for industry, but also will

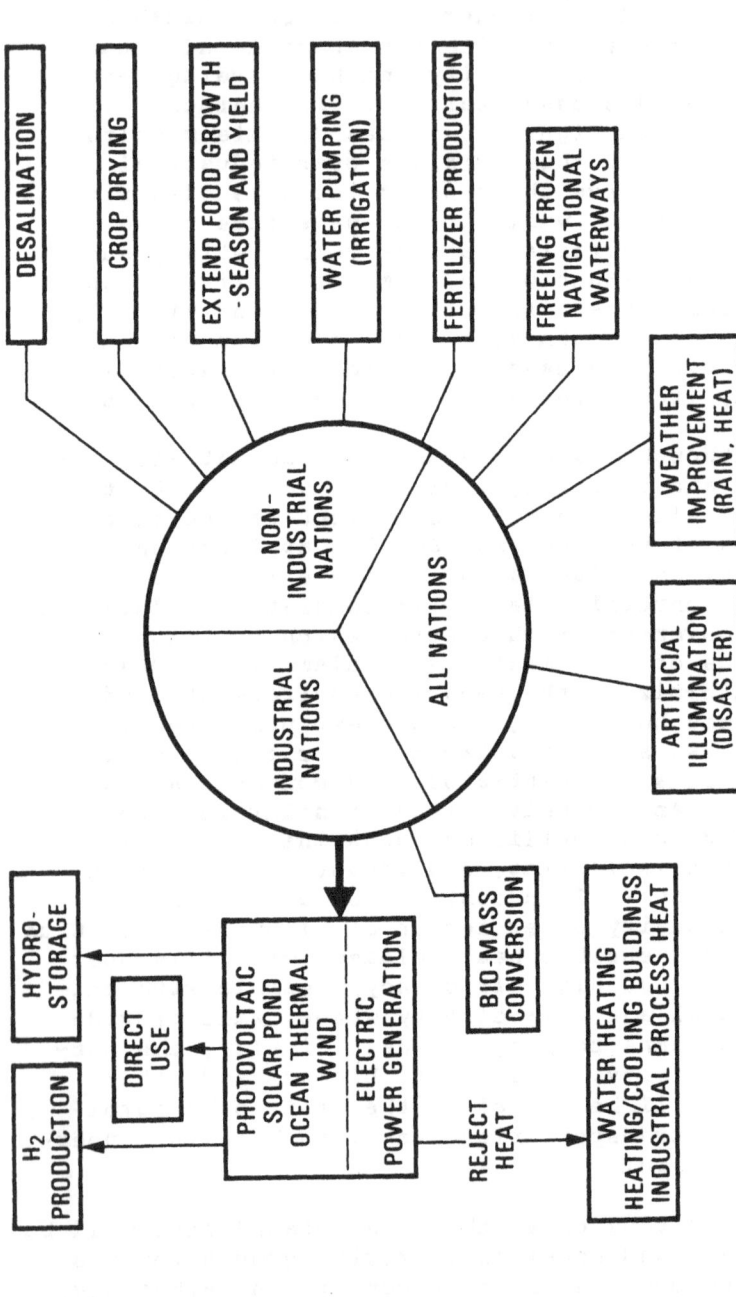

Figure 9. Multiple and simultaneous use possibilities for the solar energy provided by SOLARES.

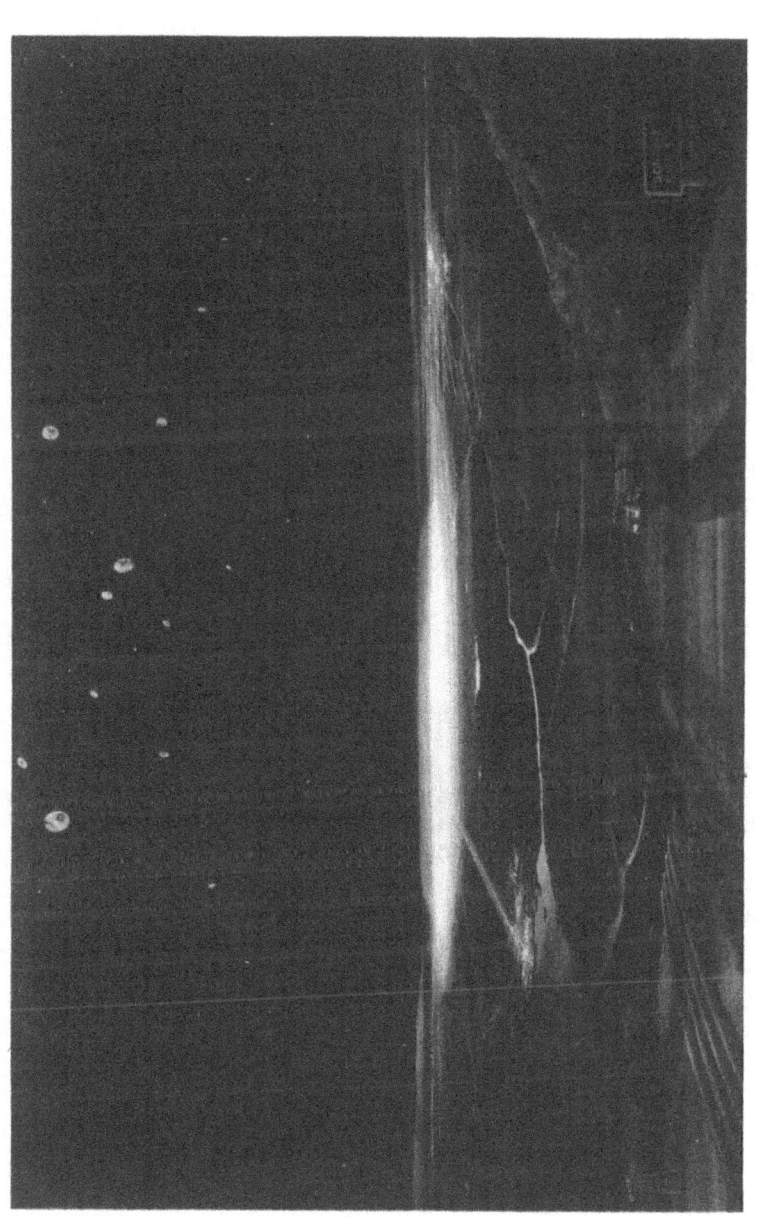

Figure 10. Artist's concept of the multiple conversion "mix" and energy use at a SOLARES conversion site. Energy and product transport is also shown.

produce many solar conversion technical advances
and cost reductions (because of the scale of a
SOLARES system) which will benefit the small, de-
centralized use of solar energy.

The SOLARES Baseline Conversion System

Without the benefit of knowing the optimum
conversion "mix" for SOLARES, a somewhat "worst
case" baseline conversion system has been assessed.
This uses horizontal "flat plate" solar cell conversion; heat dispersal (rather than economic use
of the "reject" heat) by once-through cooling,
electric storage to provide baseload power, and
other elements such as power conditioning which
constitute a solar photovoltaic conversion system.[7]
Conservative (no technology "breakthrough") costing
for the 1985-1990 time frame has been assumed.
Thus, solar cells are at $500/peak kW, in agreement
with the D.O.E. 1986 Goal and near term (battery)
electric storage at $36/kWh with 75% depth of dis-
charge giving an effective cost of $48 kWh, etc.,
are assumed. These, together with the insolation
profiles, determine the baseline system costs per
unit area of terrestrial farm for each orbital
altitude. As seen in the 4146 km summary shown in
Table 2, solar cell costs predominate; storage
costs are of more significance for the lower, more
highly eclipsed, orbits. Also shown are the
assumed efficiencies and loss factors which again,
with one exception, are those standardly used. The
exception is a plant capacity factor, or fractional
time of delivery of rated plant output to total
available time. This is assumed to be CF= 0.95 in
contrast to the standard power industry baseload
system value of 0.70. The latter number arises
from planned outages (ca. 10%), unscheduled outages
(ca. 10%) and system reserve capacity which exceeds
routine demand (ca. 10%). Because of the scale of
a SOLARES site, however, it is believed that a
highly modularized system will be necessary. Fortunately, this is consistent with a photovoltaic
power plant but, of course, is not the case for
conventional power plants. Routine, automated
maintenance, and the storage capacity reserve
should allow the SOLARES CF assumed.

An additional, important CF consideration is
cloud obscuration. Proper site location usually
reduces this to less than 20% for most studies of

photovoltaic conversion farms. However, it is our current belief that a controlled, correct sensible heat release from a SOLARES farm will possibly prevent cloud impingement or formation over the site. Gemini satellite photographs dramatically show such a naturally-occurring process at work for Guadelupe Island, located off the coast of Baja California, Mexico. Clouds moving toward the island from the Pacific Ocean are seen to be deflected around the plume of solar-heated air rising from the island. This eliminates obscuration for a site of roughly the same dimensions as a SOLARES site. An important technology challenge to future SOLARES study will be the large scale modeling and computation necessary to understand and make use of these and other possible weather modifications associated with SOLARES.

The Fuel Production Alternative

Recent study has begun to examine the efficacy of fuel production to complement the baselined electric-only system discussed above. Such SOLARES energy output is attractive for many important reasons. Since at many conversion sites electric output would greatly exceed those traditionally connected to the grid, interfacing problems would be reduced if some SOLARES energy were put into chemical form and transported by other means. Secondly, the choice of site location is greatly expanded; equatorial sites, consistent with the lowest cost mirror system and the possibility of high solar concentration and resultant temperature, would be allowed since their energy could be supertanked in the form of methanol or ammonia to points of usage. The environmental impact of SOLARES may also be lessened in this way. Fourth, it must be appreciated that fuels represent an even larger need, by approximately a factor of 4, than electricity for even the industrialized nations. SOLARES might answer this need while avoiding balance of payments deficits and conserving our fossil resources. Finally, such artificial fuel production may decrease the CO_2 buildup in our atmosphere by using atmospheric CO_2 as a chemical feedstock in the synthesis process.

The most desirable means for SOLARES fuel production would be by direct, photo-stimulated reactions. This research area has expanded greatly

Table 2. Elements of the baselined photovoltaic conversion system.

| | Sizing | Base cost | Efficiency, % | 4146 km orbit case | |
				Cost/area, $/m²	Fraction, %
Solar cells (DOE 1986 goal cells)	hα-circle	$500/pk kW	15	75	65
Storage (Battery)	Yearly peak to meet demand curve	$35.7/kWeh	75	16.67	14
Power conditioning	Yearly peak of demand curve	$50/kWe	85	7.80	7
Heat dispersal (once-through)	Adequate to meet yearly peak at ΔT = 20°C	CWP:$1/m² Pumps: $0.04/m² H. Ex. $4.96/m²	––	6.29	5
Structure, site preparation and wiring	hα-circle	$10/m²	––	10	9
			Total cost/array area	$115.76	100

Table 2. (continued)

Efficiencies: Photovoltaic array	$\eta_{PVA} = \eta_{PV}\eta_A\eta_W\eta_T = (0.15)(0.95)(0.90)(1.0) = 0.1283$
Overall System	$\eta_{SYS} = \eta_{PC}\eta_{PVA} = (0.85)(0.1283) = 0.1091$

Capacity factor: CF = 0.95
Total site size: $h\alpha$-circle + buffer = 30 km (19 mile) radius
Number of sites possible: 6
Total site cost: $135.5 B
Yearly energy output/site = 1.18×10^{12} kWeh
Average power output/site = 135 GWe
Cost/rated output power[a] = $1304/kWe

[a]Includes R&D, facilities, hardware, 15% contingencies, and interest during construction at 8%/annum.

recently and will possibly evolve means, coupled
with the economic advantage of SOLARES' continuous
radiation, to be of significant importance. A
second conversion approach would be thermal. Finally, the more expensive method, but one of present
engineering and economic predictability, is that
of solar-to-electric conversion followed by the use
of electric power for chemical production.

The production of two chemicals: ammonia and
methanol by the latter method have been examined.
Each uses circa 1985 predicted achievement in high
pressure electrolysis of water to obtain the hydrogen feedstock necessary for synthesis. Although it
has been frequently argued that hydrogen could be
directly transported for fuel usage, the expense
and safety problems associated with cryogenic
supertanker transport are avoided with methanol,
and, to a lesser extent, ammonia. A high tempera-
ture/pressure catalyzed reaction of hydrogen with
atmospheric nitrogen produces ammonia. Dugger
et al.,[8] have evaluated this process and its economics. Catalytic combination of the hydrogen with
CO_2, obtained from the atmosphere or by steam-
stripping from the carbonates in sea water, produces methanol as examined by Steinberg and
Dang.[9,10]

Significantly, the additional SOLARES plant
capital investment necessary for such fuel synthesis is only 5-15% of the baseline electric production system. This results in quite acceptable
costs of about $130/ton for ammonia (currently
costing about $150/ton using cheap natural gas as
feedstock) and $17/bbl for methanol (currently
costing about $40/bbl).

Significant and Renewable Energy -- The SOLARES Promise

An important measure of the worthiness of
pursuing a new energy alternative is its potential
impact upon future energy need. A second consideration is its ability to do so perpetually; i.e.,
using a non-depletable or renewable energy resource.
The rigorous insolation calculations, together with
an assumed terrestrial conversion efficiency,
allow the SOLARES impact or promise to be assessed.

A particularly useful way to summarize this impact for various possible orbital altitudes is to consider the following system buildup scenario. Mirrors are continuously added to the system and their altitude raised while four conversion sites simultaneously are continually expanded in radius. At the beginning, the modular sites operate on ambient solar energy to accrue revenues. After 1.6 year, the ground sites have reached the size necessary and the mirrors, at 2109 km altitude, are sufficient to allow operation at an average 1.14 kW/m^2 radiant intensity. Continuing this process, a 4146-km system would be obtained at 6 years. Finally, if desired, further expansion to a 6384-km system would be attained at 15 years.

The impact on world energy needs of such a scenario would be significant. Figure 11 compares the radiant energy/year falling on the four SOLARES sites with the predicted world energy needs.[11] Depending upon the degree with which the SOLARES energy can be converted to use, by the proper "mix" of solar conversion and "reject" heat technologies that are evolved, the potential for such a system to supply a large fraction of world energy can be appreciated. Two U.S.-owned sites, if we allow islands or other ocean bases, may be possible. Additionally, realizing that the U.S. uses about one-fourth of the present world consumption, this study allows us to appreciate the potential impact of the total four-site output of this SOLARES system on future global needs.

If we make the conservative baseline case assumption, however, of only converting the energy to electrical output and with overall system conversion efficiency of 11%, then the effect of one site on the predicted U.S. consumption[12] is seen in Figure 12. With advanced solar cells projected to be available in the 1990's[13], the higher impact shown would result.

Economic Feasibility --
The SOLARES Challenge and Promise

The economic challenge to the SOLARES concept is, as discussed earlier, to provide sufficient economic improvement to solar energy conversion so that it attains competitiveness with alternate fossil and nuclear generation. In order to

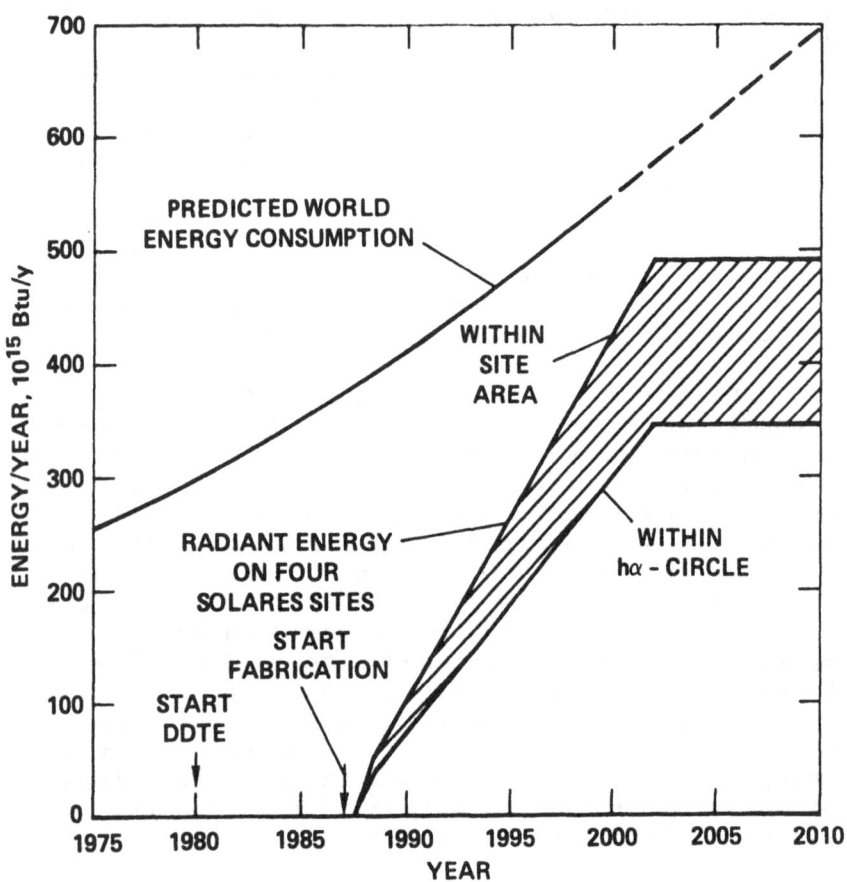

Figure 11. A comparison of the predicted world energy consumption with the radiant energy which would be supplied by a SOLARES system expanded as described in the text.

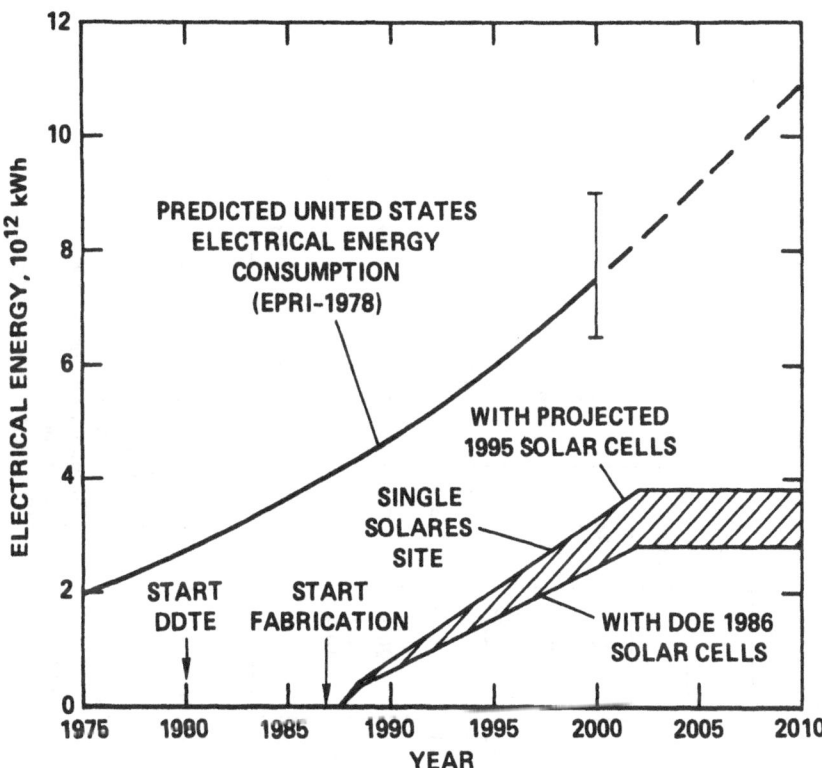

Figure 12. Potential impact of the electric output from a single SOLARES site on predicted U.S. needs. (Uniform growth scenario, as described in the text.)

Table 3. Selected baseload generation candidate alternatives – mid 1990's. (In 1977 $)

Plant type	Direct coal burning[g]	Nuclear[h]	Combined cycle coal liquid[i]	Fluid bed combustion[l]	Gas boiler – advanced gasifier[l]	Combustion turbine – advanced gasifier[l]	Advanced fuel cell advanced gasifier[l]
Capital cost ($/kW)[a]	700	830	345	800[k]	795[k]	730[m]	730[k]
Capacity factor (%)	70	70	70	70	70	70	70
Average heat rate (Btu/kWh)	10200	10400	8100	9800	12200	9100	9100
1995 O&M costs							
Fixed ($/kW-y)	2.53	2.86	1.20	10.99	10.93	14.39	13.59
Variable (mills/kWh)[b]	4.16	0.72	1.27	5.26	1.86	2.11	2.02
1995 fuel cost ($10^6/Btu)	1.11	.777	3.75	1.11	1.11	1.11	1.11
Levelized capital cost (mills/kWh)[c]	17.1	20.3	8.4	16.5	19.5	17.9	17.9
Levelized O&M cost (mills/kWh)[d]	8.6	2.2	2.8	13.3	6.9	8.4	8.0
Levelized fuel cost (mills/kWh)[e]	23.7	19.9	59.8	22.7	28.3	21.1	21.1
Levelized busbar energy cost (1977 mills/kWh)[f]	49.4	42.4	71.0	52.6	54.6	47.4	47.0

Data source: EPRI Technical Assessment Guide, EPRI-PS-866-SR, June 1978

[a] Includes all direct costs, 15% for contingencies, interest during construction at 8%/annum.
[b] Includes consumables.
[c] 15% fixed charge rate, 30 yr at 6% annual inflation.
[d] 30 yr at 6% annual inflation.
[e] Escalation rate as specified in Technical Assessment Guide.
[f] 15% fixed charge rate.
[g] Plant has flue gas desulfurization.
[h] 0.2% tails, no fuel reprocessing.
[i] Plant burns coal-derived oil.
[j] Plant burns coal.
[k] Technology design/cost goal.
[l] Plants operate on synthetic gas produced on site from coal.
[m] Preliminary process design cost.

examine this possibility, it will be necessary to
establish the predicted energy costs for these alternatives in the 1990's -- the possible SOLARES
implementation period. Assumption of a baseline
SOLARES system and current best-estimate, and
frequently non-optimized, costing for it will then
allow a similar "baseline" energy cost for SOLARES.
Comparison between these will then allow a means to
examine the sensitivity of SOLARES' competitiveness
to uncertain cost elements, such as space costing,
and to certain technology options, such as mirror
size.

It is particularly important to examine concepts, such as SOLARES, which have not yet had
in-depth systems and economic analyses in this
fashion because it minimizes the importance of sub-
jective and tentative costing while maximizing in-
sight into essential economic viability of the concept in comparison with its competition.

Busbar Costs of the Alternatives

The accepted measure of economic comparison
between alternative electric power systems which
serve the same function, such as baseload output,
is the busbar energy cost, levelized over the
operating lifetime of the system.[14] The levelized
busbar costing methodology, as elucidated by Doane,
et al.,[15] determines that representative price per
unit of energy which, if held constant throughout
the life of the system would provide the required
revenue, assuming that all cash flow interim
requirements or excesses are borrowed or invested
at the utility's rate of return. The algorithm
used to determine the $\overline{BBEC}$ correctly adjusts all
costs, expressed in current dollars, for inflation,
and in some required cases such as for fuel costs,
for escalation. Proper accounting of the total
revenue needs, both capital and interest, during
construction are assessed. Unfortunately, space
will not permit a full discussion of this well-
documented[15] method of levelized busbar energy
cost, $\overline{BBEC}$, calculation.

Table 3 summarizes the key elements used, and
the calculated values of $\overline{BBEC}$ for the most probable baseload generation alternatives (as listed
by EPRI) for commencement of operation in the mid-
1990's. All costs are expressed in 1977 dollars.

Table 4.   SOLARES baseline system.

Configuration:	SPACE SYSTEM 4146 km inclined orbit, 45,800 km² total mirror area
	GROUND SYSTEM 6 sites with DOE 1986 goal solar cells @ 15% efficiency, 11% overall system conversion efficiency, *hα*-circle area = 1,168 km² each, 135 GWe each
Impact:	Total system would produce 3.24 times current U.S. consumption, total area = 84 × 84 km² (52 × 52 mi²)

Baselined costs (in 1977 dollars):

IMPLEMENTATION SCHEDULE
5 yr development, design, test and evaluation (DDTE)
2 yr manufacturing and transport fleet facilities preparation
6 yr space and ground hardware construction

System complete about 1995

DIRECT COSTS ESTIMATE ($ B)

Facilities	$ 47.30	
Hardware	885.65	
Total direct		$932.95

INDIRECT COSTS ESTIMATE ($ B)

15% contingency on direct costs	$139.94	
Design, development, test and evaluation	43.80	
Interest[a]		
Facilities	23.58	
Hardware	101.26	
DDTE	41.01	
Total indirect		349.59

TOTAL COST	$1282.54

INDIRECT COST FACTOR	1.38	
INSTALLED COST PER RATED OUTPUT ($/kWe)[b]		1508
CAPACITY FACTOR (%)		95
1995 O&M COSTS		
Fixed ($/kW-y)		3
Variable (mills/kWh)		2
LEVELIZED CAPITAL COST (mills/kWh)[c]		27.2
LEVELIZED O&M COST (mills/kWh)[d]		4.5
LEVELIZED BUSBAR ENERGY COST (mills/kWh)[e]		31.6

Table 4. (continued)

COMPARISON BASELOAD POWER SYSTEMS (CIRCA 1995):
    Conventional coal/nuclear mix[f]

Levelized busbar energy cost (mills/kWh)[e]	45
Ambient sunlight photovoltaic[f,g]	
Levelized busbar energy cost (mills/kWh)	115

---

[a]4% first yr, 8%/annum until positive cash flow after year 11.

[b]Includes all direct costs, 15% contingency, interest during implementation at 8%/annum.

[c]15% fixed charge rate, 30 yr at 6% annual inflation.

[d]30 yr at 6% annual inflation.

[e]15% fixed charge rate.

[f]See text; these do not include their historically extensive R&D costs which are included, however, in SOLARES costing.

[g]Uses same terrestrial costing algorithm as SOLARES which results in indirect cost factor of 1.37.

Input data for the plants was obtained from the
EPRI Technical Assessment Guide of June 1978 [14] and
from DeMeo, et al.[7] It should be noted that the
EPRI capital cost per installed kW estimates shown
may appear low in comparison with recent press
accounts of certain potential installations, such
as the Sun Desert nuclear facility in California,
estimated at $1700/kW, the coal power plant at Four
Corners at $1120/kW, and a nuclear plant in the
Philippine Islands at $1700/kW. However, we will
use the EPRI costs as the representative "standards
of the industry," being mindful however, that environmental and social factors relegating against
the burning of coal or the use of nuclear energy
may ultimately increase these values.

It is seen that the $\overline{BBEC}$ values for coal, synthetic coal-derived oil, and nuclear generation
range between 42-71 (1977) mills/kWh. We will use
a value of 45 mills/kWh, which is representative
of a mix of coal and nuclear generation, as that
value which SOLARES must attain to achieve economic
competitiveness with these alternatives.

To begin this sensitivity analysis, the $\overline{BBEC}$
for the baseline SOLARES system is assessed. The
4146 km, inclined orbit case which services six
sites, three each spaced along 30°N and 30°S latitude, is chosen as the baseline system. The principal cost elements, as seen in Table 4, are the
R&D, Facilities (including the space construction
facility for assumed mirror fabrication rather than
deployment), and Space and Terrestrial Hardware
(mirrors and solar photovoltaic conversion farms).
As discussed previously, space-related costs were
derived mainly from related studies for the SPS,[5]
the Outlook for Space,[16] and studies which analyze
space manufacturing.[17] The baseline mirror has
$\pi/4$ km^2 area and mass density σ_b=8.76 gm/m^2. The
unit installed cost for this highly replicative
structure, which would be produced in large numbers
(45,800/($\pi/4$) $\simeq$ 58,000 needed) has been assessed at
$2M. This represents a unit mass cost of $291/kg.
The baseline photovoltaic solar conversion costs,
discussed earlier and presented in Table 2, are
known with greater certainty. For an hα-circle
area of 1,168 km^2 (451 mi^2), a unit site cost of
$135 B results. Each site has a power rating of
135GWe. Finally, the recent history of power plant

construction has shown numerous cost overruns. To allow for this, a contingency allowance of 15% has been included. Of course, design, development, test and evaluation has also been included as an indirect cost. Another important indirect cost, interest on capital during implementation, greatly increases the above. For the rate of expenditure shown in Table 4, the total, overall indirect cost factor then becomes 1.38. This results in a baseline system SOLARES installed cost per rated power output of $1508/kWe.

The resultant baseline system SOLARES BBEC is seen to be 31.6 (1977) mills/kWh, as compared with the alternative mean value of 45 mills/kWh. Another comparison of interest is that of the busbar cost for a normal solar farm, operating on sunlight alone, but with sufficient storage to provide baseload output. If costing identical with that used for the SOLARES ground sites is used, a BBEC of 115 (1977) mills/kWh results. The clear promise of the SOLARES concept is seen: a solar-electric energy cost reduction of a factor of about 3.6.

Sensitivities to Assumed SOLARES Costs

It is especially informative to examine the allowable range of certain key SOLARES costs which are consistent with the competitive BBEC of 45 mills/kWh. Because of the relative uncertainty of SOLARES space costs, and the overriding importance of the highest net cost component of SOLARES -- the solar cells -- these parameters are examined first, as shown in Figure 13. The solid curves indicate allowable solar cell costs per peak kW and efficiencies for various increases in the assumed baseline space cost. For reference, the DOE 1986 goal of $500/pk kW is shown as a band since this goal does not specify the cell efficiency. Since it is generally accepted to be for silicon cells, a conservative AMI value would be 15%. This, of course, is the cell assumed in the baseline SOLARES costing. As seen, a SOLARES space cost increase by a factor of 4 over that baselined (such as due to increased transportation costs, increased mirror density, etc.) could be tolerated and the SOLARES BBEC would remain competitive to the fossil/nuclear alternative.

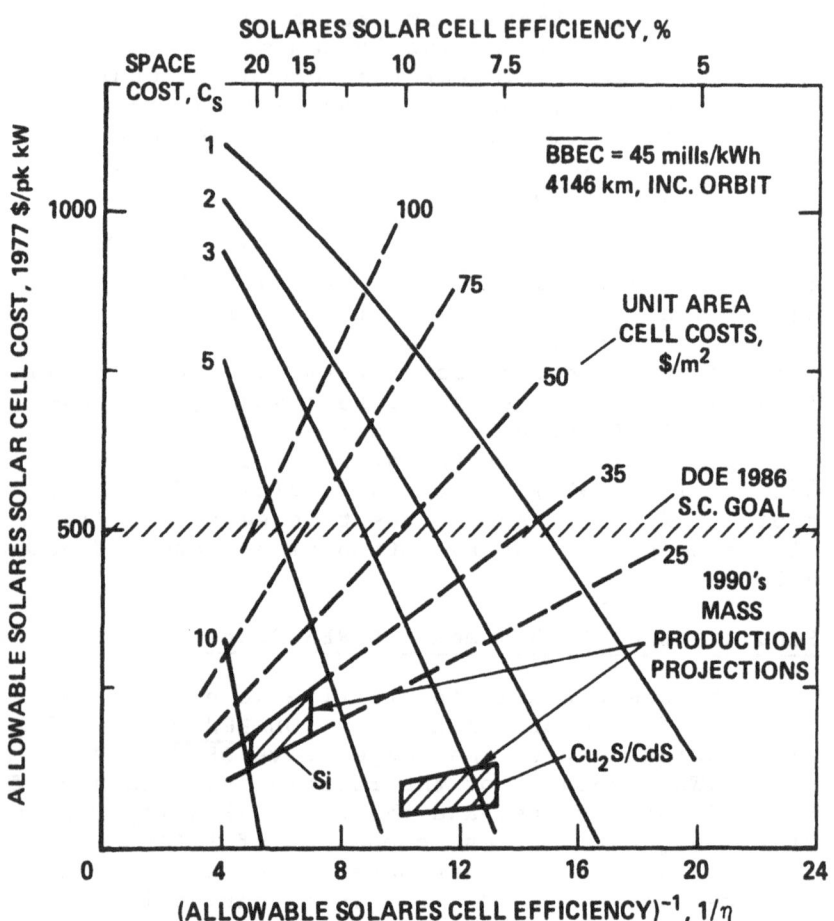

Figure 13.   Allowable solar cell cost per peak kW
and efficiencies for baselined space cost, C_S and
greater, for a 1995 competitive levelized busbar
energy cost of 45 (1977) mills/kWh.

As is seen, this is also approximately the case for the projections[7] which have been made on lower efficiency, spray-deposited cells such as Cu_2S/CdS which would appear desirable for SOLARES use. Finally, recent study by Woodcock and others[13] on the cost reductions which should obtain for cell high-production rates (somewhat <u>lower</u> than those needed for SOLARES) and taking into account the base costs associated with material and energy requirements for silicon cells, predicts circa 1990 costs of $25-35/m^2$. As seen, these would allow a space cost increase of about 10 and still provide a $\overline{BBEC}$ of 45 mills/kWh.

Such insensitivity of SOLARES energy cost to the most uncertain cost element is, of course, particularly comforting. A second, and similar analysis examines its sensitivity to the mirror areal density. This parameter, which, of course, is related to the material and transport costs of the mirrors, is important to estimation of allowable unit mirror area and, hence, the total number of mirrors needed in the space system. As discussed earlier, with our present Mark I design, approximately 50 km^2 mirrors would require $\simeq 20$ σ_b. As seen in Figure 14, such would be allowed with the advanced cells predicted by Woodcock.[13] This would reduce the baseline system mirror number to approximately 916 and greatly simplify the logistics of the system. The more conservative cell goals are seen to allow 100%, or mirrors of about 30 km^2. Advanced mirror design will hopefully even improve these numbers.

These sensitivity studies show that large errors in space costing, or alternatively, large reductions in the number of mirrors necessary, are allowable with SOLARES while yet maintaining a competitive economic posture with alternative circa 1990 fossil and nuclear power sources. They help to lend general credence to the promise of SOLARES

Social/Political Acceptability-- The SOLARES Challenge

In addition to the three criteria previously discussed, the ultimate acceptance of a new energy option, and indeed, the continuation of old ones, is closely tied to the social/political desirability of the system. Are the inevitable negative

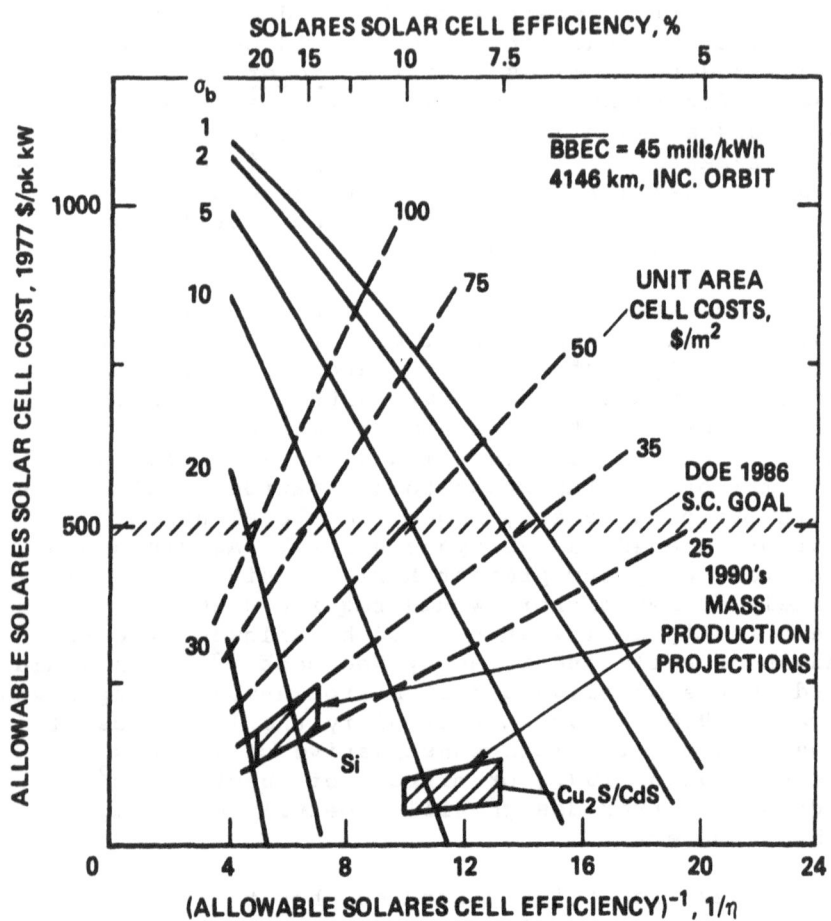

Figure 14.   Allowable solar cell cost per peak kW
and efficiencies for baselined areal mirror density,
σ_b, and greater, for a 1995 competitive levelized
energy cost of 45 (1977) mills/kWh.

environmental impacts tolerable and sufficiently
outweighed by the positive impacts? Is the economic cost -- in dollars -- and the social cost -- in
committed man years -- acceptable? Will the system
promote domestic and world stability or will it
disrupt it and, in fact, thus be a fragile target
for terrorist or military disruption? Finally, and
of course ultimately, is the system really perceived as necessary, and if so, is it the best
choice between other options?

Complete answers to these questions in relation to SOLARES are, of course, beyond the scope
of this paper. However, certain facts are clear
and will be briefly discussed.

Environmental Impact

As with any system of the magnitude of SOLARES
capable of generating a large fraction of the
world's energy needs, environmental impact is inevitable. Those which have been identified are
seen in Table 5, listed in positive, negative, and
uncertain categories. Clearly, a positive impact
will be to allow conservation of non-renewable
fossil fuels which are currently used for electric
power generation. In addition, SOLARES power may
also be used for other applications such as transportation, for which fossil fuels are now the only
economical option. Additional conservation would
accrue if the solar farm reject heat were used for
such purposes as space heating, pumping of fluids,
water desalination, industrial process heat and
crop drying. Significantly, in addition to fossil
fuel conservation, SOLARES would also eliminate the
chemical and thermal pollution due to the extraction, transporting, and burning of these fuels or
similar problems existing with nuclear fuels.

Limited investigation of the potential photo-
chemical effect of the added SOLARES solar radiation in the atmosphere shows very little net dis-
turbance; in fact, a slight positive or negative
change in ozone production, depending upon the uv
reflectance cutoff of the mirrors, was found by
numerical simulation.

The question of weather modification must
await large scale numerical computation on weather
models which contain the important fluid physics

Table 5.   Identified SOLARES environmental impacts.

POSITIVE

- Conserve fossil fuels for
  non-electrical needs

- Use electricity for *new* uses
  now only provided by fossil fuels

These decrease chemical
pollution due to mining,
transporting and burning
fossil fuels

UNCERTAIN

- Photochemistry

- Weather modification

NEGATIVE

- Land usage
- Light glint and scattering
- Space transportation pollution
- Sensible heat release

and the scales of the SOLARES system. Unques-
tionably, the possibility of weather improvement is
potentially one of the most important uses for an
orbiting mirror system. Simple, but important,
applications such as removal of fog from critical
transportation areas (airports, harbors, etc.),
prevention of killing frost to certain expensive
crops, extension of growing seasons, and snow
cover removal, appear possible. More subtle
changes may be effected by the addition of the
SOLARES energy with the proper timing and in the
proper global location. These possibilities must
await future detailed and very complete study.

The potential negative impacts are also listed in Table 5. Generally, they await full technical assessment and with this, hopefully, technology "fixes." Historically, the land usage associated with extensive solar-electric production has
been cited as a fault of this energy option. However, because SOLARES provides a five-fold increase
in average energy per unit area over that of normal sunlight, this is not the case here. In fact,
SOLARES land usage, including a buffer zone of an
extent to include all reflected insolation down to
0.1 of the central intensity, would be a factor of
4.4 less than normal solar farms and would approximately equal, per unit of power generated, that of
fossil and nuclear plants. For the baseline system
which would exceed present-day world electrical
consumption, a total area of only 138 X 138 km^2
(86 X 86 mi^2) would be needed.

Without further study and optimization, the
remaining impacts are more difficult to assess.
One is the visual effect of light scatter from the
mirrors and from the site. The former would be of
particular concern to astronomers. The latter
would be similar to the general glow seen above a
desert city at night, and observable out to about
150 km. Most scattering above the site is "small
angle" and originates from particulates below 3 km
altitude. Because of the scale of the site, virtually all of the scattered radiation will fall on
the site. It should be appreciated that in all
scattering situations conceived, either from a
single mirror which inadvertently sweeps a reflected sun image across the ground during an orientational maneuver (called "glint") or in the cases
mentioned above, the intensity of the light is very

small and could produce only a visual effect. For
reference, it must be remembered that the full,
collective intensity of all the mirrors above the
UVC is only that present in the desert at noon.
The chemical and noise pollution of the launch
vehicles necessary for implementation must be minimized. Finally, as discussed previously, limits
for sensible heat release at the site must be established. These will enable the design of a conversion "mix" which will minimize undesirable
weather modification by SOLARES.

Social-Economic Impact

Clearly, the implementation of a new energy
source which effectively replicates much of what
now exists would have profound effects upon the
society which produces it. Men, material, and
financial resources would be required in a measure
which has few parallels in peace-time endeavor. In
a recent interesting paper,[18] Vajk, et al., examine
the various means for the management and financing
of a system (the SPS) of similar magnitude. Reasonable methods, involving combinations of governmental, corporate, and public institutions, do
appear possible.

To place the SOLARES financial needs into perspective, we recall that the baseline system cost
was about $5.9 B/yr for the five years of research
and development, $17.6 B/yr for two years of facil-
ties preparation and $185 B/yr for six years of
space and terrestrial hardware production. This
sums to $1,173 B or an average yearly value of
$90 B(excluding interest). For comparison, this
value is less than yearly monies collected into the
U.S. social security system; it represents about
20% of the federal budget; it is roughly 5% of the
U.S. GNP; it is only three times the estimated
annual investment of the U.S. electric power
industry needed to just keep up with increases in
U.S. demand. With the inclusion of other bene-
fiting countries into the supply of capital, if
desired, the cost of a significant shift to a
solar energy economy would, therefore, appear
reasonable. Alternatively, with sole U.S. financing at a higher burden, the sale of SOLARES insolation would again establish this country as a
major exporter of energy. Gross revenues from the
baseline system would be $262 B per year at

37 mills/kWh, 15% rate of return. At 45 mills/kWh, they would be nearly $400 B/yr.

The Minimal Cost System

It should be noted that the baseline system does not represent the lowest capital investment SOLARES system. Although a full study of the modified space costing for the hardware of a low altitude (1177 km) equatorial orbit has not yet been completed, it is clear that it will be much smaller than the baseline since it requires only one-seventh as many mirrors. Higher requisite mirror areal density removes some of this advantage, however. The estimated cost (including identical R&D and facilities costs) would be $148 B for mirrors plus one ground site. Each additional 10.4 GWe site would cost $12.6 B; a maximum of ten could be serviced. Such a system would find its primary use in fuel production, as discussed earlier, and its yearly output would be equivalent to about 10^9 bbl of oil. The present U.S. oil import is about 3×10^9 bbl/yr. Thus, even this "Minimal Startup SOLARES System" would significantly help to reduce the approximately $110 M/day contribution of oil to the trade deficit.

Can We Afford "Free" Energy? -- The Interstate Highway System Analogy

As has been shown, the cost of the baselined system is very large, in consonance with its potential energy impact. A crucial concern is the availability of this capital -- the "front end money" -- necessary to implement it and thus allow the 30-yr return on investment which the competitive busbar cost would give. There are many possible ways to reduce this initial investment, such as beginning with the aforementioned minimal system and later expansion. Another is to recognize the potential importance of this solar energy option to the country's needs and to support much of the research and development from general funds. This, of course, has been extensively done in the area of nuclear fission reactor research for the past 35 years, and hence, these tax supported costs do not appear in the unit cost of $830/kW quoted earlier. Since much of the SOLARES R&D would be of multiple benefit, this approach appears justified.

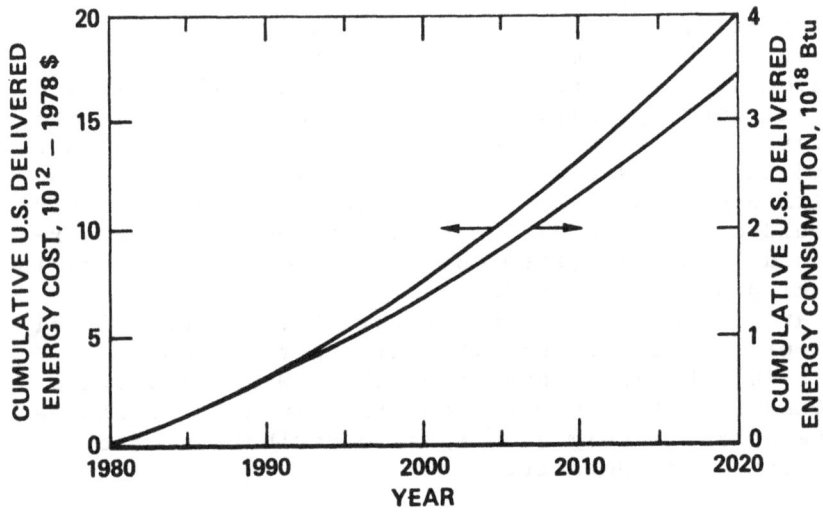

Figure 15.   Projected cumulative U.S. consumption
and expenditure for delivered energy from the
year 1980.

There is an interesting analogy between financing a new energy source like SOLARES and that of
the country's interstate highway system. For both
military and economic reasons, the taxpayers and
their representatives decided to accrue revenues
which would result in the future "free" use of
these roads. The total cost of the 42,500 mile
system will be about $170 B in 1978 dollars; it is
presently 92% complete. These revenues have been
collected over a 25-year period by taxes on motor
fuel (4¢/g), commercial vehicles and parts, tires,
oil, etc. -- i.e., they are a "potential user tax."
Thus, there is precedent for the accumulation of
very large capital sums in an equitable way from
those who will benefit by its use.

A similar method could be used for SOLARES, or
other new energy source, which would be based upon
the concept of an energy replacement tax: those who
consume energy (mostly non-renewable fossil-derived)
would be assessed a tax per unit of energy used
which would pay for the development of the means to
provide energy for future generations. Although
the full implications to the national economy of
such a tax are very complex and have not been
assessed, an examination of Figure 15 shows its
capital gathering potential. The figure shows the
predicted cumulative delivered energy consumption
and cost to the U.S. users, as calculated from data
given in a recent SRI study.[19] For example, during
the 1980-1995 period, 983 x 10^{15} Btu of energy will
be consumed at a total cost of 5,270 X 10^9 1978
dollars. A 1% surtax would thus give $53 B; 5%
giving $264 B. If the period were extended to 2025,
that is, the end of a nominal "book life" of a
power system beginning operation in 1995, these
numbers would be $235 B and $1,175 B, respectively.
In comparison, the estimated (baseline) cost of all
(both space and terrestrial) research and development, DDTE, and the space components of SOLARES
is $84 B, exclusive of interest. Each of the six
ground conversion farms would add $158 B, for a
total of $948 B.

From these numbers, without benefit of extended economic analysis of the proper mix of investments, spending, interest on borrowed money, etc.,
which such an energy replacement fund (or "international energyway system" fund, in analogy with
the interstate highway system fund) would follow,

it is estimated that capital in sufficient quantity
could be available to support SOLARES. A 1-3% tax
would allow the fund to provide the mirror system
and then sell the SOLARES insolation to the entities (public, private, or international) which
finance the ground sites; a 5-7% tax might allow
two of the sites to be purchased for domestic
"free energy" as well.

The "sale" or exportation of the radiant
energy supplied by SOLARES is a concept which
raises a number of interesting points. Foremost
is "what is its value?" An approximate answer can
be obtained by comparing the levelized busbar cost
of energy which a SOLARES plant could competitively
charge, 45 mills/kWh as discussed earlier, with
that which would necessarily be charged to simply
obtain return of capital ground site investment and
pay operations and maintenance. With a unit cost
(terrestrial cost only) of $1304/kW, the latter
number is 25 mills/kWh. Thus, the value of SOLARES
radiation, essentially "fuel" to the plant owner,
is 45-25=20 mills/kWh. Charging such a rate would
return levelized revenues of about $150 B/yr to the
energy trust fund -- a revenue which would allow
eventual mirror system expansion. Since at least
four sites would probably be foreign-owned, energy
export to these would also help alleviate a balance
of trade deficit (currently about $15 B/yr).

Again, it must be emphasized that these numbers are only preliminary estimates. However, they
indicate one method, related to the replacement of
energy used, which appears to be available for
accumulating the monies necessary for a new world
energy system such as SOLARES.

Conclusion

The SOLARES concept has been examined under
the four criteria which any candidate energy system
must satisfy. Our assessment of the technical
feasibility indicates that although there are many
challenges which remain for further detailed study
and resolution, no "showstoppers" have been found
which would demand other than reasonable extension
of current space technology and terrestrial solar
conversion methods. Recent, independent investigation by others on low mass density mirrors of the
SOLARES genre, which may be deployable and scaled

to larger areas with lower mass than our design, is
particularly encouraging. It is believed that similar progress will be made in the solar conversion
area when the opportunities afforded by the nearly
continuous, higher average radiant intensity of
SOLARES is fully appreciated. In this way, the
economically and environmentally desirable mix of
conversion methods, rather than the expensive,
electricity-only method of photovoltaic conversion
which was used here to expeditiously "baseline" a
system, will result. The second criterion examined, significant and renewable energy impact,
showed the real promise of SOLARES to supply energy
to the world community. If the system were expanded to the size discussed, the radiant energy
supplied would exceed present, total world energy
use. Our ability to evolve a high conversion
efficiency or total-use system at the sites will
determine the ultimate utility of this SOLARES
promise. Significantly, even with the low efficiency of the baselined system, a single, highly
modular site would provide about 40% of the predicted U.S. electrical needs in year 2002. The
third criterion examined, economic feasibility, was
approached in a manner consistent with our present
lack of in-depth systems and economic assessments.
A baselined system, with space and terrestrial
costs consistent with studies by others, was compared with the circa 1995 candidate baseload power
systems listed by EPRI. It was found that with
projected 1986 DOE solar cells, and certainly with
more advanced cells predicted by others, large
overruns in the most uncertain cost element of
SOLARES -- the space component -- would not change
the competitive posture of SOLARES-derived energy
to the fossil/nuclear mix alternative. On the
other hand, if the baselined costs could be
achieved, greater technical flexibility results --
such as the use of higher mass, very large mirrors
which would significantly reduce the number of
satellites required. Finally, a brief examination
of the fourth criterion, social/political acceptability, showed the many questions yet remaining
in this area. The positive impacts of conserving
fossil chemicals and thereby avoiding their pollution in power generation or transportation use are
believed important.

However, the most significant unanswered
question about SOLARES -- that of possible weather

modification -- must be resolved by future study.
There is hope that this will show how SOLARES, or
some variant of it, could provide beneficial
weather enhancement. Certainly, the studies will
also define the allowable sensible heat release
density in the vicinity of the conversion sites and
thus prescribe the conversion and dispersal goals
which must be achieved. Finally, the question of
the ability to finance this bold step into large-
scale, solar-derived energy for the world was
examined. One particularly attractive means, the
assessment of an energy replacement tax, was estimated to be capable of developing the SOLARES space
system for a levy of 1-3% on domestic delivered
energy. This investment would again allow our
nation to regain its position of energy exporter,
providing revenues in excess of $100 B/yr. A 5-7%
levy would probably be sufficient to buy two sites
and hence provide "free energy" for the populace
in analogy to their "free" use of the tax-purchased
interstate highway system. Putting such low cost
energy into the domestic grid would, it appears,
not only improve the quality of life directly, but
also benefit the economic posture of our commercial
entities in the world market place.

References

1. Billman, K.W.; Gilbreath, W.P.; and Bowen,S.W.; Introductory Assessment of Orbiting Reflectors for Terrestrial Power Generation, NASA TM X-73, 230, April 1977.

2. Billman, K.W.; Gilbreath, W.P.; and Bowen, S.W.; Satellite Mirror Systems for Providing Terrestrial Power: System Concept, The Industrialization of Space, vol. 36, part I, Advances in the Astronautical Sciences, AAS, San Diego, CA, April 1978, pp.391-414.

3. Billman, K.W.; Gilbreath, W.P.; and Bowen,S.W.; Orbiting Mirrors for Terrestrial Energy Supply, Radiation Energy Conversion in Space, K.W. Billman, ed., AIAA Progress in Astronautics and Aeronautics Series, vol. 61, July 1978, pp. 61-80.

4. Private communication, 2 November 1978.

5. Piland, R.O.: The Solar Power Satellite Concept Evaluation Program, Radiation Energy Conversion in Space, K.W. Billman, ed., AIAA Progress in Astronautics and Aeronautics Series, vol. 61, July 1978, pp.3-24.

6. Friedman, L., et al.: Solar Sailing, the Concept Made Realistic, Paper 78-82, AIAA 16th Aero-space Sciences Meeting, Huntsville, AL, January 1978.

7. DeMeo, E.A.; and Bos, P.B.: Perspectives on Utility Central Station Photovoltaic Applications, Solar Energy, vol. 21, 1978, pp. 177-192.

8. Dugger, G.L.; Francis, E.J.; and Avery, W.H.; Technical and Economic Feasibility of Ocean Thermal Energy Conversion, Solar Energy, vol. 20, 1978, pp. 259-274.

9. Steinberg, M.; and Dang, V.; Production of Synthetic Methanol From Air and Water Using Controlled Thermonuclear Reactor Power-I. Technology and Energy Requirement, Energy Conversion, vol. 17, 1977, pp. 97-112.

10. Dang, V.; and Steinberg, M.; Production of
    Synthetic Methanol From Air and Water Using
    Controlled Thermonuclear Reactor Power --II.
    Capital Investment and Production Costs,
    Energy Conversion, vol. 17, 1977, pp.133-140.

11. Energy Facts II, Science Policy Research Divi-
    sion, Congressional Research Service, Library
    of Congress, U.S. Government Printing Office,
    August 1975.

12. Williams, L.J.; Boyd, J.W.; and Crow, R.T.;
    Demand 77, EPRI Special Report, EA-621-SR,
    vol. 1, March 1978.

13. Woodcock, G.R.: SPS Cost Considerations, J.
    Energy, vol. 2, July-August 1978, pp.196-202.

14. Technical Assessment Guide, EPRI Technical
    Assessment Group, EPRI Special Report,
    PS-866-SR, June 1978.

15. Doane, J.W., et al.: The Cost of Energy from
    Utility-Owned Solar Electric Systems, JPL
    Report JPL 5040-29, Jet Propulsion Laboratory,
    Pasadena, CA, June 1976.

16. Anon.: A Forecast of Space Technology 1980-
    2000, NASA SP-387, Scientific and Technical
    Information Office, Washington, DC, 1976.

17. Davis, E.E.; and Miller, K.H.: Construction of
    a 10 GWe Solar Power Satellite, 13th Inter-
    society Energy Conversion Engineering Confer-
    ence Proceedings, SAE, Warrendale, PA, 1978,
    pp. 189-194.

18. Vajk, J.P.; Stutzke, R.D.; Salkeld, R.; Stine,
    G.H.; and Klan, M.S.: Satellite Power System
    (SPS) White Paper on Financial/Management
    Scenarios, Science Applications, Inc.,
    Pleasanton, CA, 10 October 1978.

19. Fuel and Energy Price Forecasts: Quantities
    and Long Term Marginal Prices, EPRI Report
    EA-433, vol. 1, September 1977.

Overview

Robot and automation technology, as outlined by Ewald Heer, are destined to play a significant role in future space missions. Increased reliance on robotics and automation offers the likelihood of enhancements of mission capabilities along with reductions in overall costs. Areas where this technology can be effectively employed include planetary exploration, global service satellites (such as LANDSAT), and space industrialization. An aggressive robot and automation research and development program in such areas as artificial intelligence, machine intelligence and robotics will greatly facilitate the design of a cost-effective and flexible space program.

16. Robots and Automation in the Space Program

Introduction

Most of those who are acquainted with the space program expect that the space transportation system including the Space Shuttle and its adjuncts will open up a new era of expanded space activities in scientific exploration and global service undertakings. The first steps towards utilizing the space environment for industrial and commercial ventures will become possible. This can trigger requirements for more advanced transportation systems in the 1990's that will enable and can lead to large-scale projects. By the end of this century, such projects could include a Space Solar Power System for solar energy production, a lunar base for extracting and processing material resources, and a space station for commercial processing and manufacturing. In addition to developing the necessary transportation systems, a major objective for the space program is the development of systems that will enable the cost effective construction and erection of such systems in space.

On examining potential space missions during the next two decades, it is evident that robot and automation technology can be a vital contributor to the cost-effective implementation and operation of both scientific exploration missions (earth orbital and planetary) and public service missions (large communication satellites and large scale systems). In some cases, it can make the mission feasible and in other cases, it can make it economically affordable.

This paper briefly examines the possible role that could be played by robots and automation in the space program. It also explores and conjectures on benefits that can be expected as a result.

Background

Looking back on industrial evolution, there are a number of key steps: division of labor, use of mechanical power, standardized parts, transfer devices, open-loop control, closed-loop control, and computer control. These steps indicate a progression in the amplification of human capabilities. Until the dawn of the industrial revolution, human physical capabilities (muscle power) were amplified through the use of simple machines. Then automation also came into the picture. After a slow beginning, automation has been moving on with ever-increasing pace, helping to accomplish more and more human tasks. By means of limit switches and governors, the machine became able to make simple decisions on its own, such as "go-stop" or "more-less". Some outstanding examples are assembly line transfer devices in automobile manufacturing during the early twentieth century, systems with open-loop mechanical control around World War I, and systems with closed-loop control approximately during World War II.

In the 1940's, the advent of the digital computer also opened up the possibility of performing many so-called low-level (and possibly high-level) intellectual activities automatically. In addition to the long-standing amplification of human physical capabilities, the amplification of the human mind was in the making.

Since the first commercial computers were introduced in the early 1950's, the "electronic brain" has brought widespread changes. It has reached into innumerable crevices of our lives. At last count, some 1800 firms and more than 4 percent of our labor force were dedicated to maintaining and extending automation through computers. Today computer-automated operations include collecting, maintaining, and assimilating enormous amounts of data. Automatons greatly simplify operations and reduce costs. They pay bills and keep track of accounts, inventories, personnel, and pricing. In factories, many routine or dangerous tasks are done by computer-controlled manipulators. Even some engineering designs and analyses are done automatically.

Not counting the hundreds of thousands of microprocessors proliferating in our society, we have today more than 200,000 free-standing computers. If one added up the performance of these machines in terms of their human equivalent, one would find that it would take more than 400 billion people to accomplish the same tasks that these machines perform today. Considering that, conservatively, at most 20 percent of the U.S. population is engaged in relevant work using these computers, we have built into our society a mind-amplifying factor of about four orders of magnitude. Clearly, America depends on automated machines so much that the country would come to a grinding halt without them. The man-machine symbiotic existence is here, and this relationship is moving with ever increasing speed in the direction of more automation.

Along with the fast pace of these developments several branches of science and technology came into being. Probably computer science emerged as the most encompassing discipline on equal par with other engineering professions such as electrical engineering and mechanical engineering. Along came also changes in the definition and use of old familiar terms. For instance, automation in manufacturing is retaining its usual connotation of self-acting and self-regulating machinery performing specialized pre-programmed tasks; however, in the business world, the words automation and computers as the central element in automated systems are only now becoming dominant, and the newly emerging robot devices are general-purpose systems with a great degree of autonomy. By means of a computer "brain" as the central element, a robot is able to sense its environment, plan and decide its actions, and perform mechanical manipulations and data handling tasks sometimes to a degree normally done by humans. The brain, consisting of computer hardware and software, incorporates the robot's intelligence, i.e., machine intelligence. The development of machine intelligence technology relies heavily on advances in the science themes of

artificial intelligence such as problem solving,
scene analysis, theorem proving and speech understanding.

Robots and Automation in NASA Planning

While mechanical power provides <u>physical
amplification</u> and computers provide <u>mind amplifi-
cation</u>, telecommunication provides <u>space amplifi-
cation</u>, i.e., amplification of the space accessible
to humans. By means of telecommunication, humans
can activate and control systems at remote places.
They can perform tasks even as far away as the
planets. During the 1960's, this became known as
teleoperation.[1] <u>Teleoperators</u> are man-machine
systems that augment and extend human sensory,
manipulative, and cognitive abilities to remote
places. In this context, the term <u>robot</u> can then
be applied to the remote system of a teleoperator
if it has at least some degree of autonomous sen-
sing, decision-making, and/or action capability.
The concept of teleoperation has profound significance in the space program. Because of the large
distances involved, almost all space missions fall
within the teleoperator definition, and because of
the resultant communication delay for many missions
the remote system requires autonomous capabilities
for effective operation.[2] The savings of operations time for deep space missions can become tremendous if the remote system is able to accomplish
its tasks with minimum ground support. For example
it has been estimated that a Mars roving vehicle
would be operative only 4 percent of the time in a
so-called move-and-wait mode of operation. With
adequate robot technology, it should be operative
at least 80 percent of the time.

Because of the central role of computers in
many space systems, both <u>robotics</u> and <u>automation</u>
have pervaded the space program, at least to some
degree, in many guises since its inception. However, these trends did not take sufficient advantage of the technological developments during the
last one or two decades. Some have criticized
NASA for this, stating that the computer technology
used in NASA missions is at least ten years behind
the times. While some of the criticisms are not
justified, some may be valid. The valid ones can
usually be traced back to the generally accepted
requirement of flawless performance of NASA

missions (often for political reasons) and to long lead times from project conception to implementation. Projects and missions are usually conceived and their objectives defined based on then proven technology rather than on the technology existing at the time of the project design period.

The development and application of robot and automation technology during the first decade of NASA's existence was hardly organized. Johnsen and Corliss[1] in 1967 presented a comprehensive survey of teleoperator and robot technology to that date. The development of teleoperator technology had its beginnings for serious, practical applications in the late 1940's and early 1950's, when it became necessary to control remotely located manipulators in hazardous environments of nuclear laboratories. With few exceptions, this work was done under the aegis of the Atomic Energy Commission. Nevertheless, in 1966, Case Institute of Technology, working under a NASA grant, demonstrated a computer-controlled manipulator that could perform pre-programmed tasks specified by the operator.

Surveyor 3 and Surveyor 7 landed on the moon in 1967 and 1968, respectively. The claw and arm mechanism of the surface sampler gave the spacecraft the capability of a crude robot. Controlled remotely from earth at each step, it dug trenches and performed soil experiments.

The first organized effort to address the questions of teleoperator/robot technology planning for the space program took place in 1970, when NASA conducted an inter-Center study on requirements for the years 1970-1990.[3] The study team recommended to the NASA Administrator a research and development program with focus on all subsystems (manipulators, mobility units, sensors, displays, controls, communications, artificial intelligence, etc.) and system integration.

Some highlights of the robot technology planning, development, and application activities during the past decade have been presented in Ref. 4.

A series of planning studies and workshops was initiated with the Outlook for Space Study in 1974[5], which included a comprehensive forecast

Figure 1. Galileo spacecraft navigates between Jupiter and Galilean satellites in rendering. After sending a probe into the Jovian atmosphere, the robot spacecraft will perform complex maneuvers at various inclinations with repeated close en- counters with the satellites.

of space technology for 1980-2000. In a subsequent NASA/OAST Space Theme Workshop[6], the technology forecasts were applied to three broad mission themes: <u>space exploration, global services and space industrialization.</u> Based on the derived requirements for cost-effective space mission operations, Kurzhals[7] identified five new directions for space electronics technology developments: (1) automated operations aimed at substantial reduction of mission support costs; (2) precision pointing and control; (3) efficient data acquisition to permit a large increase in information collection needed for global coverage; (4) real-time data management; and (5) low-cost data distribution. The machine intelligence and automation technologies for data acquisition, data processing, information extraction and decision making emerge here as the major drivers in each area and call for the systematic exploitation of the spectacular developments in electronics during the past two decades. In addition, for certain areas such as automated operations in space, the mechanical technologies directed at materials and objects acquisition, handling, and assembly must also be further developed; robots doing construction work in earth orbit or on the lunar surface will need manipulative and locomotion devices to perform the necessary transport and handling operations.

Future Applications

In space applications, robots may take on many forms. None looks like the popular science fiction conception of a mechanical man. Their appearance follows strictly functional lines, satisfying the requirements of the mission objectives to be accomplished.

Space Exploration

Space exploration robots may be exploring space from earth orbit like orbiting telescopes, or they may be planetary flyby and/or orbiting spacecraft like the Mariner and Pioneer families. They may be stationary landers with or without manipulators like the Surveyor and the Viking spacecraft, or they may be wheeled like the Lunakhod and the proposed Mars rovers. Others may be penetrators, flyers, or balloons and some may bring science samples back to earth (Figs. 1-3). All have in

Figure 2. Mars surface robot will operate for
2 years and travel about 1000 km performing experiments automatically and sending the scientific
information back to Earth.

Figure 3.   Artist's concept of a Mars surface
scientific processing and sample return facility.
Airplanes transport samples into the vicinity of
the processing station.   Tethered small rovers then
bring the samples to the station for appropriate
analysis and return to Earth.

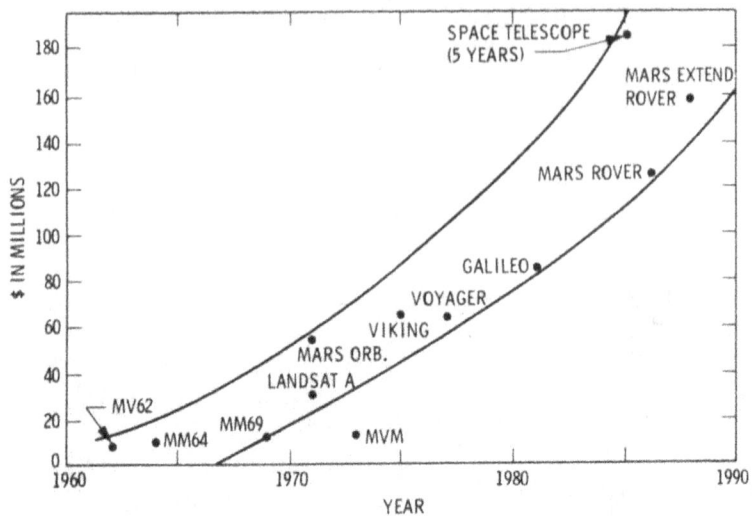

Figure 4. Trend of mission ground operation costs.
Increasing mission complexity and duration contribute to the ground operation costs.

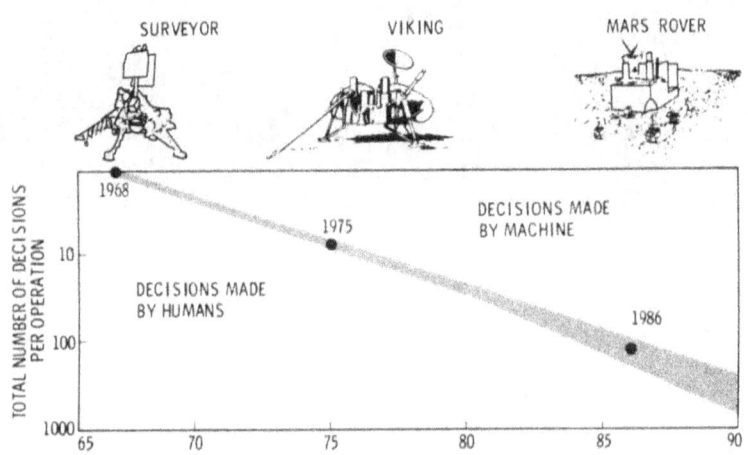

Figure 5. Trend of function (decision) allocation
between humans and robots. For the same typical
operation, the robot takes over an increasing
number of elementary functions, leaving the high-
level decisions to humans.

common that they can acquire scientific and engineering data using their sensors, process the data with their computers, plan and make decisions, and send some of the data back to earth. Some robots are, in addition, able to propel themselves safely to different places and to use actuators, manipulators, and tools to acquire samples, prepare them, and perform in situ experiments or bring them back to earth.

Exploration robots are required to send back most of the collected scientific data, unless it becomes repetitive or is meaningless and can be discarded. The unknown space environment accessible to the sensors is translated into a different, still uninterpreted environment, in the form of computer data banks on earth. These data banks are then accessible for scientific investigations long after the space mission is over.

As on-board data acquisition requirements and mission complexity increase, more data is sent to the ground that must be coordinated and handled. If mission durations also increase, the ground-based mission operations will quickly use up a major part of the total mission cost (Fig. 4). With more capable on-board machine intelligence, the robot will be able to perform many of the ground-based operations in space autonomously which would relieve ground activities and would greatly contribute to cost reductions for the same mission operations. Adapting advanced automation techniques to ground-based operations would also lead to significant cost savings. The combined effect of applying advanced automation technology to exploration missions is projected to yield a cost improvement of about two orders of magnitude by the year 2000.

A qualitative measure for the level of machine intelligence is the number of decisions made by the robot versus those made by its human supervisor. For a particular operation like "dig a trench", the total number of elementary decisions is presumably constant at approximately 1000. With increased machine intelligence capabilities, the robot will make a correspondingly increased number of the required decisions (Fig. 5).

Projections into the future lead one to speculate on the possibility of highly autonomous exploration robots in space. Such exploration robots would communicate to earth only when contacted or when a significant event occurs and requires immediate attention on earth. Otherwise, they would collect the data, make appropriate selective decision, achieve it, and store it on board. They would serve as a data bank, and their computers would be remotely operated by accessing and programming them from earth whenever the communication link to the robot spacecraft is open. Scientists would be able to interact with the robot by remote terminal. Indeed, the concept of distributed computer systems, presently under investigation in many places including JPL, could provide to each instrument its own microcomputer, and scientists could communicate with their respective instruments. They could perform special data processing on board and request the data to be communicated to them in the form desired. Alternatively, they could retrieve particular segments of raw data and perform the required manipulations in their own facilities on earth.

Prime elements in the "scientist-deep space robot" link would be large antenna relay stations in geosynchronous orbit. These stations would also provide data handling and archiving services, especially for exploration robots which will not be accessible after a relatively short time, e.g., those leaving the solar system.

Global Services

Global services robots orbit the earth. They differ from exploration robots primarily in the intended use of the collected data. They collect data for public service use on soil conditions, sea states, global crop conditions, weather, geology, disasters, etc. These robots generally acquire and process an immense amount of data. However, only a fraction of the data is information of interest to the ultimate user. At the same time, the user often likes to have the information shortly after it has been obtained by the spacecraft. For instance, the value of weather information is short-lived except for possible historical reasons. The value of information on disasters such as forest fires is of even shorter duration. The demand for high-volume on-board data processing and pertinent automated information extraction is therefore great.

The usual purpose of global services robots is to collect time-dependent data on the earth's environment, whose static properties are well known. The data is used to determine specific patterns or classes of characteristics and translate these into useful information. For instance, for LANDSAT and Seasat A (Fig. 6), the data is presently sent to the ground, where it is processed, reduced, annotated, analyzed, and distributed to the user. This process requires up to three months for a fully processed satellite image and costs several thousand dollars. The image must then be interpreted by the receiver, i.e., the information must still be extracted by the user.

Present developments in artificial intelligence, machine intelligence and robotics suggest that in the future, the ground-based data processing and information extraction functions will be performed on board the robot spacecraft. Only the useful information would be sent to the ground and distributed to the users, while most of the collected data could be discarded immediately. This would require the robot to be able to decide what data must be retained and how it was to be processed to provide the user with the desired information. For instance, the robot's computer brain could have a large number of pattern classification templates stored in its memory or introduced by a user with a particular purpose in mind. These templates would represent the characteristics of objects and/or features of interest. The computer would constantly compare the scanned patterns with those stored in its memory. As soon as something of interest appeared, it would "take a closer look" by zooming in on the pattern and examining it with higher resolution, comparing it to a progressively narrower class of templates until recognition had been established to a sufficient degree of confidence. The robot would then contact the appropriate ground station and report its findings and, if required, provide the user with an annotated printout or image. One can envision that the user would be able to interact with the robot, indeed with his particular instrument, by remote terminal, much the same as with a

Figure 6.   Seasat A.   The oceanographic satellite's high-data-rate Synthetic Aperture Radar imaging device has provided data on ocean waves, coastal regions, and sea ice.

central computer and, depending on intermediate results, modify subsequent processing.

For space exploration and global services, the ground-based mission operations can become an incredibly complex process. For planetary missions, these mission operations are handled primarily by the Deep Space Network and the Space Flight Operations Facility at JPL.

The latest example of planetary exploration mission, and perhaps the most complex to date, is Viking. At times there were several hundred people involved in data analysis, mission planning, spacecraft monitoring, command sequence generation, data archiving, data distribution, and simulation. While for earlier space missions sequencing had been determined in advance, on Viking this was done adaptively during the mission. The operational system was designed so that major changes in the mission needed to be defined about 16 days before the flight action. Minor changes could be made as late as 12 hours before sending a command. The turnaround time of about 16 days and the number of people involved contributes, of course, to sharply increased operational costs (Fig. 4). The Viking operations costs are for a 3-month mission. The planned Mars surface rover mission is expected to last two years, covering many new sites on the Martian surface. Considering that this mission would be more complex and eight times as long, ground operations would have to be at least ten times as efficient to stay within, or close to, the same relative costs as for Viking.

During the Viking mission, about 75,000 reels of image data tapes were collected and stored in many separate locations. The images are now identifiable only by the time when and the location where they were taken. No indication regarding image information content is provided, and the user will have to scan catalogs of pictures to find what he wants. For such reasons, it is expected that most of the data will not be used again.

The ground operations for earth orbital missions suffer from similar problems [7] as those for planetary missions. The overall data stream is usually much higher here, images are still very costly, and they take up to several months to reach the user.

These considerations lead one to conclude that technology must be developed so that most ground operation activities can be performed as close as possible to the sensors where the data is collected, namely by the robot in space. However, examining the various ground operations in detail, it becomes clear that most of those that must remain on the ground could also be automated with advanced machine intelligence techniques. The expected benefits derived from this would be a cost reduction for ground operations of at least an order of magnitude and up to three orders of magnitude for user-ready image information.

Utilization of Space Systems

Space industrialization requires a broader spectrum of robotics and automation capabilities than those identified for space exploration and global services. The multitude of systems and widely varying activities envisioned in space until the end of this century will require the development of space robot and automation technology on a broad scale. It is here that robot and automation technology will have its greatest economic impact. The systems under consideration range from large antennas and processing and manufacturing stations in earth orbit to lunar bases, to manned space stations, to satellite power systems of up to 100 km^2. These systems are not matched in size by anything on earth. Their construction and subsequent maintenance will require technologies not yet in use for similar operations on earth.

Space processing requires a sophisticated technology. First, it must be developed and perfected, and then it must be transferred into the commercial arena. Basic types of processes presently envisioned are: solidification of melts without convection or sedimentation, processing of molten samples without containers, diffusion in liquids and vapors, and electrophoretic separation of biological substances. It is expected that specialized automated instrumentation will be developed for remote control once the particulars of these processes are worked out and the pressure of commercial requirements becomes noticeable.

Large-area systems such as large space antennas, satellite power systems, and space stations
require large-scale and complex construction facilities in space. Relatively small systems, up to
100 m in extent, may be deployable and can be
transported into orbit with one Shuttle load. For
intermediate systems of several-hundred-meter
extension, it becomes practical to shuttle the
structural elements into space and assemble them on
site (Figs. 7-9).

Very large systems require heavy-lift launch
vehicles which will bring bulk material to a construction platform (Fig. 10), where the structural
components are manufactured using specialized
automated machines (Fig. 11).

The structural elements will be handled by
teleoperators or self-acting cranes and manipulators which bring the components into place and join
them (Fig. 12). Free-flying robots will transport
the structural entities between the Shuttle or the
fabrication site and their final destination and
connect them. These operations require a sophisticated general-purpose handling capability. In
addition to transporting structural elements, the
robot must have manipulators to handle them, and
work with them and on them. Large structural
subsystems must be moved from place to place and
attached to each other. This usually requires
rendezvous, stationkeeping, and docking operations
at several points simultaneously and with high
precision -- a problem area still not investigated
for zero gravity. Automated (smart) tools would
also be required by astronauts to perform special
local tasks.

These robot systems could be controlled
remotely like teleoperator devices, or they could
be under supervisory control with intermittent
human operator involvement. Astronauts in space
or human operators on earth will need the tools to
accomplish the envisioned programs. The technology for in-space assembly and construction will
provide the foundation for the development of
these space-age tools.

After the system has been constructed, its
subsequent operation will require service functions
that should be performed by free-flying robots or

Figure 7.   Large space systems require robot and
automation technology for fabrication, assembly and
construction in space.

Figure 8.   Large space antennas are erected with
the help of a space-based construction platform.
The Shuttle brings the structural elements to the
platform, where automatic manipulator modules under
remote control perform the assembly.

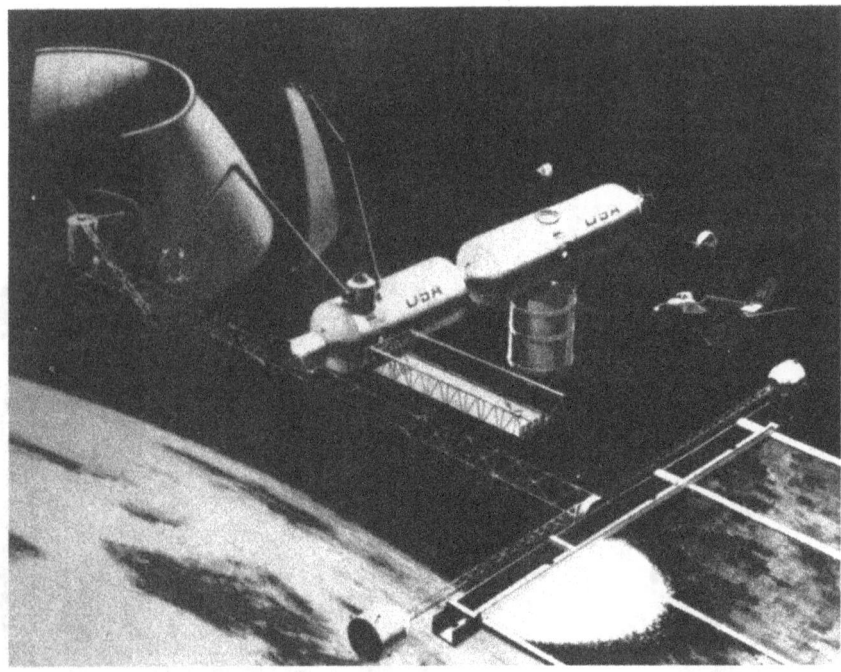

Figure 9. Construction of a space station. Bulk
material is brought by the Shuttle. Structural
elements are fabricated at the construction facility and then assembled by remotely controlled
manipulators.

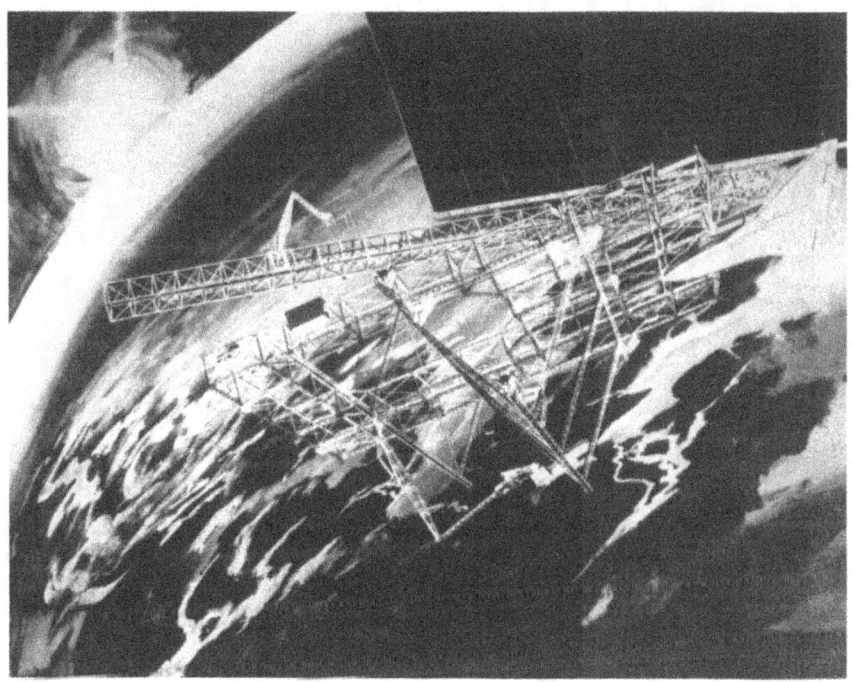

Figure 10.   Complex construction facility in space
with automatic beam builders, cranes, manipulators,
etc., is served by the Shuttle.

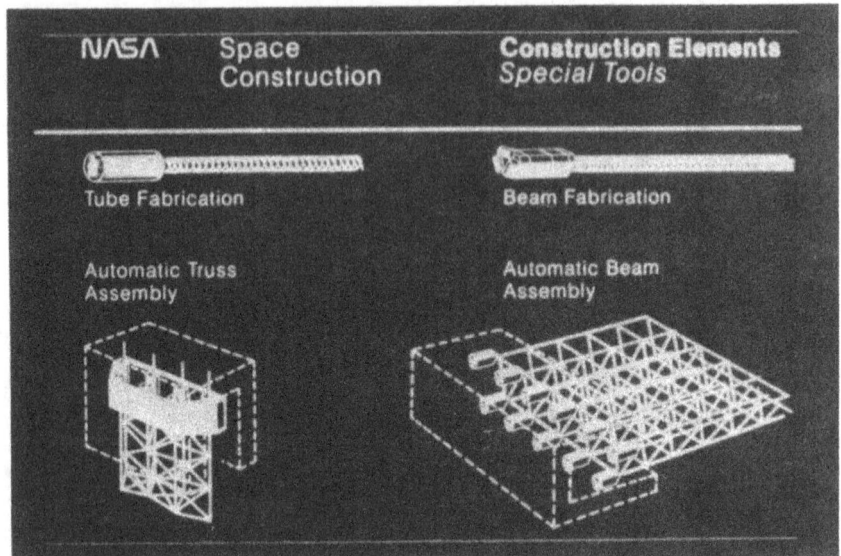

Figure 11.   Automatic beam builders use sheet metal or bulk composite materials supplied by the Shuttle.

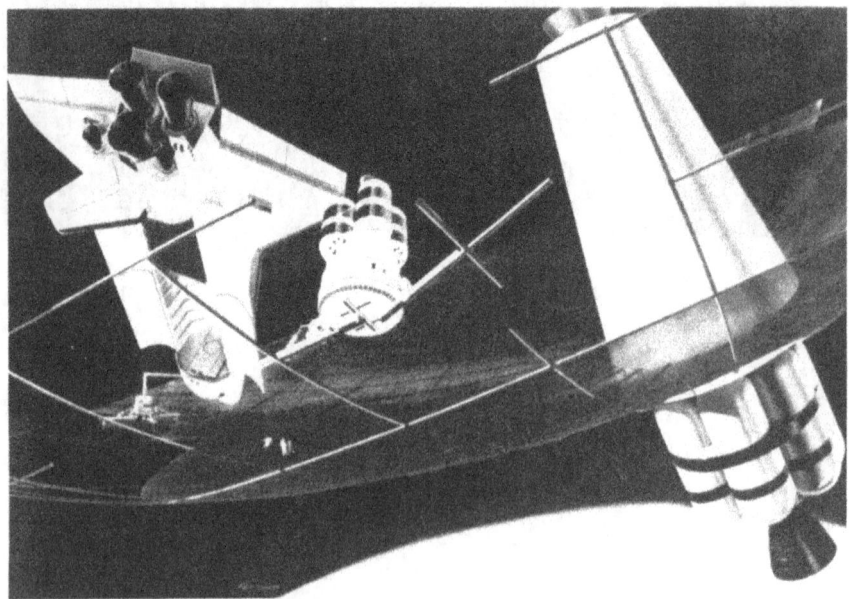

Figure 12.   Space construction of large antenna systems with automated tools, teleoperated manipulators, and free-flying robots.

by robots attached to the structure. The functions
which such a robot should be able to perform include calibration, checkout, data retrieval, resupply, maintenance, repair, replacement of parts,
cargo and crew transfer, and recovery of spacecraft.

During and after construction, there should be
a robot on stand-by for rescue operations. An
astronaut drifting into space could be brought back
by a free-flying robot. Such devices could also
be on stand-by alert on the ground. The delivery
systems for these rescue robots need not be man-
rated. They can deliver expendable life support
systems or encapsulate the astronaut in a life
support environment for return to a shuttle, space
station, or earth. They could also perform first-
aid functions.

Another phase of space industrialization calls
for a lunar base. After a lunar surface survey
with robot (rover) vehicles, an automated precursor
processor system could be placed on the moon. This
system would collect solar energy and use it in
experimental, automated physical/chemical processes
for extracting volatiles, oxygen, metals and glass
from lunar soil delivered by automated rovers
(Fig. 13). The products would be stored, slowly
building up stockpiles in preparation for a lunar
base. The lunar base would be constructed using
automated equipment and robots similarly as in
earth orbit. After construction, general-purpose
robot devices would be necessary for maintenance
and repair operations. In addition, the lunar
base would use all types of industrial automation
(qualified for operation in space) that is generally used on earth for similar tasks.

In addition to acting as an enabling catalyst
for some missions and reducing the cost of assembly
and construction in space for others, the primary
purpose of developing and applying robot and automation technology is the anticipated reduction of
operation costs for all missions. It is, therefore,
of interest to examine the NASA expenditures in
various areas and to estimate potential cost
savings, should an aggressive robot and automation
research and development program be initiated. In
a recent JPL study, it has been estimated that

Figure 13.   Automated material processors on the
lunar surface are serviced by robot vehicles with
raw lunar soil.

about 600 million of the NASA budget in 1978
dollars could be saved annually by the year 2000
for ground operations, orbital operations, and data
analysis. This does not include additional savings
for incorporating advanced automation and machine
intelligence techniques in design and testing operations. These projected savings also do not allow
for the necessary expenditures to develop the
required technology. Rough estimates for such
investments range between 200 and 250 million in
1978 dollars to be spent on research and development between 1980 and 1995. This would be on the
average about 0.4 percent of the NASA budget or
about 12 percent of the OAST space budget. Although the estimated potential savings are subject
to error, they indicate that space robot and automation technology is addressing high-leverage areas
within the space program with large possible returns on the investment.

Conclusions

In summary, the space program is entering an
era in which tremendous advances in sensor and
electronics technology offer great opportunities
to introduce advanced autonomous operations leading
to large cost savings and mission capability enhancements. The possibility of applying robot
and automation technology exists in almost every
activity within the space program. It is anticipated that the space program will continue to be
pursued with vigor during the coming decades, and
it is therefore expected that the necessary tools
will be available and that the opportunities
afforded by electronics technology will be seized
by the national space efforts in order to develop
the required robot and automation technology that
will facilitate a more capable and cost-effective
space program.

References

1. Johnsen, E.G., and Corliss, W.R., <u>Teleopera-</u><u>tors and Human Augmentation,</u> NASA SP-5047, December 1967.

2. Heer, E., "Remotely Manned Systems for Operation and Exploration in Space," <u>First CISM-</u><u>IFT.MM Symposium on Theory and Practice of</u> <u>Robots and Manipulators,</u> 5-8 September 1973, Springer Verlag, Udine, 1974.

3. Deutsch, S., ed., "Teleoperator/Robot Development Task Team Report to the Acting Administrator, NASA," October 13, 1970.

4. Heer, E., "New Luster for Space Robots and Automation," <u>Astronautics and Aeronautics,</u> September, 1978.

5. <u>Outlook for Space,</u> NASA SP-386, January 1976.

6. <u>OAST Space Theme Workshop, Final Report,</u> Langley Research Center, Hampton, VA, April, 1976.

7. Kurzhals, P.R., "New Directions in Space Electronics," <u>Astronautics and Aeronautics,</u> February, 1977.

Suggested Readings

Key Economic Areas in Chinese History as Revealed in the Development of Public Works for Water Control, Ch'ao-Ting Chi, Paragon, New York, 1963.

Science and Civilization in China, Joseph Needham et al., Cambridge University Press, 1971 (esp. Vol. 4, Part III, Civil Engineering & Nautics).

The Pyramids of Egypt, I.E.S. Edwards, Penguin, Baltimore, 1961 (revised ed.).

Technology in the Ancient World, Henry Hodges, Alfred A. Knopf, New York, 1970.

The Ancient Engineers, L. Sprague De Camp, Doubleday, New York, 1963.

The Roads That Led to Rome, Victor W. von Hagen, World, New York, 1967.

A Social History of Engineering, W.H.G. Armytage, Faber & Faber, London, 1961.

A Thousand Years of London Bridge, C.W. Shepherd, John Baker, London, 1971.

The Cathedral Builders, Jean Gimpel, Grove, New York, 1961.

Sir Hugh Myddelton, Entrepreneur and Engineer, J.W. Gough, Oxford, 1964.

Edit Du Roi Pour La Construction D'un Canal De Communication Des Deux Mers, Océane & Méditerranée, St. Germain-en-Lay, 1666.

Lives of the Engineers, Samuel Smiles, John Murray, London, 1861.

Isambard Kingdon Brunel, L.T.C. Rolt, Longmans, London, 1957.

The Great Iron Ship, James Dugan, Harper, New York, 1953.

World Ditch, The Making of the Suez Canal, John Marlowe, Macmillan, New York, 1964.

The First Transatlantic Cable, Adele Gutman Nathan, Random House, New York, 1959.

The Nile Reservoir Dam at Assuan and After, Sir William Willcocks, E&FN Spon, Ltd., London, 1903.

Public Works, Vol. I, July-October, 1903, The St. Bride's Press, London.

The Path Between the Seas, David McCullough, Simon & Schuster, New York, 1974.

Of Dikes and Windmills, Peter Spier, Doubleday, New York, 1969.

The Tallest Tower, Eiffel and the Bell Epoque, Joseph Harriss, Houghton Mifflin, Boston, 1975.

The Bridge, Gay Talese, Harper & Row, New York, 1964.

The Big E, Edward P. Stafford, Random House, New York, 1962.

Cape to Cairo, Mark Strage, Harcourt Brace, New York, 1973.

The Great Iron Trail, The Story of the First Transcontinental Railroad, Robert W. Howard, Bonanza Books, New York, 1962.

The Impossible Railway, The Building of the Canadian Pacific, Pierre Berton, Alfred A. Knopf, New York, 1972.

The St. Lawrence Seaway, William R. Willoughby, University of Wisconsin Press, 1961.

A History of Dams, Norman Smith, Citadel Press, New Jersey, 1972.

The Great Wall of France, The Triumph of the Maginot Line, Putnam's, New York, 1961.

A Hole in the Bottom of the Sea, Willard Bascom, Doubleday, New York, 1961.

The Tunnel Under the Channel, Thomas Whiteside, Rupert Hart-Davis, London, 1962.

Supership, Noel Mostert, Alfred A. Knopf, New York, 1974.

The Hindenburg, Michael M. Mooney, Dodd Mead, New York, 1972.

The Ultimate Migration, R.H. Goddard, manuscript dated January 14, 1918, the Goddard Biblio Log, Friends of the Goddard Library, November 11, 1972.

Ethics and Bigness, edited by Harold Lasswell and Harlan Cleveland, Harper, New York, 1962.

Small is Beautiful, E.F. Schumacher, Harper & Row, New York, 1973.

The Spoils of Progress: Environmental Pollution in the Soviet Union, M.E. Goldman, M.I.T. Press, 1972.

A Wild Plan for South America's Wilds, "Fortune", December 1967, p. 148.

Global Earth-Shapers, "Fortune", April 1970, p. 78.

Managing the U.S. Supersonic Transport Program (A and B), ICH #9-675-049 and ICH #9-675-050; LIQUECOAL, ICH #9-678-046 (and other case studies available through the Intercollegiate Case Clearing House, Soldiers Field Road, Boston, MA 02163).

Power From the Sun: Its Future, Peter E. Glaser, "Science", Vol. 162, November 1968, pp. 857-886.

The Colonization of Space, G.K. O'Neill, "Physics Today", September 1974, pp. 32-42.

A Space Utilization Authority (SUA) Could be the TVA of the Future, Penelope G. Grodzka, Paper presented at the Conference: "Southeast 2001: The Next 25 Years," 12 and 13 November 1976, Urban Life Center of Georgia State University, Atlanta.

Implications of Low-Cost International Non-Voice Communications, Ithiel de Sola Pool and Arthur B. Corte, Report to the U.S. Department of Commerce, September, 1975.

"Limits to the Management of Large, Complex Systems," in Assessments of Future National and International Problem Areas, Vol. II, February 1977, Stanford Research Institute, Center for the Study of Social Policy, Menlo Park, California.

"Constraints on Large-Scale Technological Projects," in Assessment of Future National and International Problem Areas, Vol. II, February 1977, Stanford Research Institute, Center for the Study of Social Policy, Menlo Park, California.

"The Apollo Tradition: An Object Lesson for the Management of Large-Scale Technological Endeavors", by Robert C. Seamans Jr., and Frederick I. Ordway. In Interdisciplinary Science Reviews, Vol. 2, No. 4, pp. 270-305, December 1977, Spectrum House, London.

The Polaris System Development, Harvey Sapolsky, Harvard University Press, 1972.

Ptolemaic Alexandria, P.M. Fraser, Oxford, 1972. (three volumes).

The Emerging Japanese Superstate, Herman Kahn, Prentice-Hall, New Jersey, 1970.

"The Tragedy of the Commons", Science, Vol. 162, pp. 1243-1248, December 13, 1968.

Decision Support Systems, Peter G.W. Keen and Michael S. Scott Morton, Addison Wesley, Reading, Mass., 1978.

The New Management, William H. Gruber and John S. Niles, McGraw Hill, New York, 1976.

The Peckham Experiment, Innes H. Pearse and Lucy H. Crocker, Yale University Press, 1945.

Macro-Engineering and the Infra-Structure of
Tomorrow, edited by Frank P. Davidson, L.J.
Giacoletto and Robert Salkeld, AAAS Selected Sym-
posium, Westview Press, Boulder, Colorado, 1978.

Index